REINVENTING DARWIN

Reinventing Darwin:

The Great Debate at the High Table of Evolutionary Theory

Niles Eldredge

John Wiley & Sons, Inc.

New York • Chichester • Brisbane • Toronto • Singapore

Copyright © 1995 by Niles Eldredge
Published by John Wiley & Sons, Inc.

Library of Congress Cataloging-in-Publication Data

Eldredge, Niles.
    Reinventing Darwin: the great debate at the high table of evolutionary theory / Niles Eldredge.
        p.   cm.
    Includes bibliographical references.
    ISBN 0-471-30301-1 (cloth : acid-free paper)
    1. Evolution (Biology)--Philosophy.   I. Title.
QH360.5.E44   1995
575′.001--dc20   94-32861

Printed in the United States of America

10 9 8 7 6 5 4 3 2 1

Contents

Prologue

The High Table

In British colleges, dining rooms traditionally feature a High Table at the head of the room, where the elite of the instructional staff, the doyens, sit aloof from the student hoi polloi. I have taken my title from a welcome back to the evolutionary "High Table" extended by the distinguished English evolutionary biologist John Maynard Smith to us paleontologists who had been laboring in the evolutionary fields for some time. I have a great deal more to say about that welcome, what prompted it, and what the nature of the discourse on evolutionary issues has been, and continues to be, around that metaphorical High Table in ensuing chapters. It is the metaphor of the High Table that concerns me most at this juncture.

For metaphor it is. There are, perhaps, a few thousand practitioners of evolutionary biology, and many more biologists, anthropologists, geologists, psychologists, and philosophers who take a deep, professional interest in one aspect or another of evolutionary biology. Evolution belongs to no one—no single individual, certainly, but also no particular discipline. The breadth of discourse is sweeping indeed, and there are very many voices contributing to the conversation.

The trouble with High Table imagery, of course, is that it is exclusionary. Readers will find the same few names cropping up repeatedly throughout. I have identified two main cadres—opposing camps—holding court around this metaphorical High Table. On one side sit the likes of John Maynard Smith, Richard Dawkins, and George Williams, main exponents of what I call "ultra-Darwinism." Sitting opposite them are people like Stephen Jay Gould, Steven M. Stanley, Elisabeth S. Vrba, and me—all of whom happen to be paleontologists, but belonging to a more eclectic group I have chosen to call the "naturalists." Many other biologists of diverse stripe and opinion show up in these pages, but the protagonists and antagonists peopling this book comprise only a small fraction of the total number of evolutionists, all with a legitimate place at the evolutionary High Table.

Most of the discussants peopling my version of the High Table have developed their views in books. Science is commonly not supposed to be conducted in books. Rather, it is supposed to be the pithy paper, communicating in terse terms the results of an experiment, a field investigation, or a computer simulation of some small portion of a greater cosmic puzzle that reports the fruits of scientific labor. The image is not altogether unapt. Scientists do write such papers, and scientific journals continue to proliferate at an amazing, even alarming, rate, simultaneously stimulating and stifling scientific communication.

Scientific articles are indeed important. My own paper, with Stephen Jay Gould, "Punctuated Equilibria: An Alternative to Phyletic Gradualism" (1972), arguably was as important an impetus behind Maynard Smith's welcoming us back to the High Table as anything else. It was, indeed, a paper, a fairly brief essay. But, on the other hand, it was published in a book, *Models in Paleobiology*, edited by Thomas J. M. Schopf.

It is a myth that scientists don't write books. Nor am I thinking of the old-age memoirs of retired wizards. Scientists have always written books that do real honest-to-goodness scientific work. Evolutionary biology is perhaps the very best example. The field was essentially founded by Charles Darwin's *On the Origin of Species by Means of Natural Selection. Or the Preser-*

vation of Favoured Races in the Struggle for Life. The long-awaited book sold out on the first day of publication: November 24, 1859.

Books have always proven to be important turning points, fulcra in evolutionary history. The "modern synthesis" in evolutionary biology owed its genesis in no small part to seminal founding documents—books—by many of its most important contributors: Ronald Fisher's *The Genetical Theory of Natural Selection* (1930), Theodosius Dobzhansky's *Genetics and the Origin of Species* (1937), Ernst Mayr's *Systematics and the Origin of Species* (1942), and George Gaylord Simpson's *Tempo and Mode in Evolution* (1944) come quickly to mind.

Nor has the pattern changed in the post-1959 era that is the focus of this present work. Ultra-Darwinism has been developed and promulgated heavily in book format, beginning with George Williams's *Adaptation and Natural Selection* (1966) and certainly including Richard Dawkins's highly popular *The Selfish Gene* (1976), to mention but two of the many books published by ultra-Darwinians in the past 30 years. Nor have we naturalists shied away from book writing, as readers of Stephen Jay Gould's works will readily attest. I also have written a number of books—all devoted, in one way or another, to the task of developing and explicating the naturalistic perspective on evolutionary biology.

The present book is designed to carry on this long-standing tradition. It is meant as a contribution to evolutionary discourse itself. But it also serves as an account, from my naturalistic perspective, of the arguments and dialogues within evolutionary biology in modern times—roughly the years following the 1959 Darwinian centennial—up to the present time. It is partisan—in the way, for example, Richard Dawkins's *The Selfish Gene* (1976) and *The Blind Watchmaker* (1986) paint a glowing picture of his side of affairs—and a rather unflattering portrait of our side of the High Table. Here, then, is our naturalistic perspective, which I find exciting, compelling, and much preferable to the inherent myopia of ultra-Darwinism.

The conversations go on and on, and it's time to listen in.

1

Setting the Table

In 1984, writing in the pages of the British scientific journal *Nature*, John Maynard Smith, England's doyen of evolutionary biology, welcomed paleontologists to the "High Table" of evolutionary discourse. Recent events, wrote Maynard Smith, called for a revision of the entrenched ". . . attitude of population geneticists to any paleontologist rash enough to offer a contribution to evolutionary theory [which] has been to tell him to go away and find another fossil, and not to bother the grownups."

With these words, Maynard Smith captured the supercilious essence, the hubris, of the inner group of geneticists who had dominated evolutionary biology at least since the late 1950s. If we paleontologists were now meant to feel welcome at the High Table, the ensuing conversations have nonetheless still been tinged by an edgy wariness on both sides. This mistrustfulness echoes through the recesses of history all the way back to 1859 with the publication of *On the Origin of Species by Means of Natural Selection*. For Charles Robert Darwin was truly the founding father of evolutionary discourse, and all sides of the basic evolutionary disputes legitimately find their patrimony in his work.

1

Why has it been so difficult for geneticists and paleontologists to establish any lasting and meaningful evolutionary dialogue? The answer transcends simple turf wars, or the inevitable misunderstandings of jargon, data, and theory of fields as disparate as genetics and paleontology. Everyone, sometimes even a paleontologist, finds it difficult to appreciate the vastness of geological time, to fathom what a million years really means in biological terms. And unless one has direct experience designing and conducting a fruit fly experiment, or has wrestled with the mathematics of a theoretical problem in population genetics, there is little chance to grasp the subtleties of changing gene frequencies in populations.

But these are not insurmountable barriers. Paleontologists really can absorb the gist of population genetics, just as geneticists have successfully grappled with the fundamentals of paleontology. Somehow, though, such ecumenism fails to clear the air completely. That the arguments continue unabated around the High Table reflects something deeper than mutual ignorance of the inner workings of foreign disciplines.

What is really at stake is diametrically opposed suppositions of how evolutionary biology should be conducted. Maynard Smith put the geneticist's position succinctly: paleontology is about history. It is the paleontologist's job to elucidate what the evolutionary process has wrought over the last 3.5 billion years.

But evolutionary *theory* is about how the evolutionary process works. And that, to a geneticist, means how genetic information, underlying the production of the physical appearance (and physiologies and behaviors) of organisms, comes to be modified over the course of time. What can a paleontologist, trapped under the dead hand of history, possibly have to say about the mechanics of genetic change?

After all, science is supposed to concern itself with how things work. As a paleontologist, I readily concede that my long-dead fossils, lacking any traces of their soft anatomies or behaviors, are totally mute on the subject of reproduction and transmission of genetic information. And that is, I acknowledge, a major limitation to our data.

But what do I see as I gaze across the High Table at my genetically imbued colleagues? I see slavish adherence to a tradition that dates back to Darwin, serving as the pivotal article of faith in a geneticist's approach to understanding evolution. The credo goes like this: we have a fundamental understanding of what causes genetic information to change within a population from one generation to the next. Factoring out random factors for a moment, that understanding is embodied in the principle of natural selection. Grasping that central truth, we then assert that evolutionary history is the outcome of natural selection working on available genetic variation.

Period. End of story. We have an elegantly simple theory of evolutionary change and, if we are to heed geneticists (and Darwin), we simply take that natural selection model of generation-by-generation change and *extrapolate* it through geological time. And that, to my paleontological eyes, is just not good enough.

Simple extrapolation does not work. I found that out back in the 1960s as I tried in vain to document examples of the kind of slow, steady directional change we all thought *ought* to be there, ever since Darwin told us that natural selection should leave precisely such a telltale signal as we collect our fossils up cliff faces. I found instead, that once species appear in the fossil record, they tend not to change very much at all. Species remain imperturbably, implacably resistant to change as a matter of course—often for millions of years.

I had rediscovered a phenomenon that turned out to be well known to Darwin's contemporary paleontologists: species are fundamentally stable entities—a phenomenon that Stephen Jay Gould and I dubbed "stasis" in our 1972 paper elaborating the evolutionary notion of "punctuated equilibria." Bringing stasis back out into the open as an evolutionary phenomenon crying out for explanation is the main reason why Maynard Smith welcomed us paleontologists to the High Table.

As shall be seen, geneticists at the High Table tend to accept stasis as a valid and most interesting evolutionary phenomenon—one meriting explanation. Meaning *their* explanation. Once duly thanked for bringing the matter to their attention, paleon-

tologists are dismissed as incapable of saying anything meaningful about stasis.

What matters most at this juncture is not the ins and outs of
the debate itself, but what it symbolizes in terms of the starkly
different approaches to evolution taken by modern paleontologists and geneticists. These are crystallized in the two main
factions currently seated around the High Table: *ultra-Darwinians*, my name for the articulators of the gene-centered and essentially reductionist approach to evolutionary explanation; and
naturalists. Naturalists include paleontologists such as myself,
but also include ecologists, systematists, and other sorts of
evolutionary-minded biologists who see the reductionist stance
of the ultra-Darwinians as a distortedly oversimplified view of
the natural world in general, and of the evolutionary process in
particular.

Who are the ultra-Darwinians and how exactly do they differ from the naturalists? The distinction that I am drawing here
corresponds in a general way to discipline boundaries—but that
alone hardly serves as a definition. Ultra-Darwinians tend to be
geneticists. But not all evolutionary-minded geneticists are ultra-
Darwinians, nor are all ultra-Darwinians geneticists. Likewise,
many paleontologists, ecologists, and systematists adopt what I
take to be the "naturalist" stance, though some paleontologists
are staunch conservatives, strongly allied with the ultra-Darwinian camp.

The real differences lie in the contrasting treatments of a
core set of basic issues—the themes and issues of this book. I
mention them here simply to set the Table and to identify the
protagonists among the dinner guests. At the very heart of the
matter lies natural selection. Ultra-Darwinians have adopted the
stance that natural selection is the central evolutionary process.
But, in so doing, they have significantly altered the very basic
conceptualization of natural selection. In a nutshell, ultra-Darwinians see natural selection as competition (among members of
the same species) for reproductive success. But that is not all.
Ultra-Darwinians see all competition, including competition for
food and other economic resources, as fundamentally an epiphe-

nomenon of the real competition: competition for reproductive success.

Richard Dawkins will probably not mind my calling him the ultimate ultra-Darwinian. It was he who went a major step further in his *The Selfish Gene*, proclaiming that it is actually genes, not organisms, that are locked in a constant, titanic competitive struggle to leave copies of themselves behind to the next generation.

When it comes to natural selection, my naturalistic perspective is markedly conservative compared to the radical ultra-Darwinian stance. I see natural selection basically as a filter. Organisms compete for resources. As a side effect of such competition, those who make out better have a somewhat greater chance of reproducing successfully—and their offspring will tend to inherit the genetic information underlying their parents' success.

I will flesh out these contrasting views on natural selection in far greater detail as the discussion unfolds. But note that ultra-Darwinians in effect have transformed the concept of selection from the traditional passive filter to a far more activist stance. To my eyes, ultra-Darwinians are guilty of a form of "physics envy." They seek to transform natural selection from a simple form of record keeping, a filter that biases the distribution of genes between generations, to a more dynamic, active force that molds and shapes organic form as time goes by.

But there is more. An ultra-Darwinian sees such active selection, such competition for reproductive success, as the underpinning of absolutely all biological phenomena. Everything flows from competition among genes, or at least among organisms, to leave more copies of genetic information to the next generation. The very existence of complex biological systems is seen to owe its genesis to such genetic competition. Sociobiology is an explicit application of ultra-Darwinian principles, asserting that social systems arise and evolve through competition for reproductive success. Even ecosystems, comprised of many different lineages (populations of different species) have not escaped the ultra-Darwinian gaze. Ecosystems, according to Dawkins, at least, are ultimately to be understood as products of competition within and among populations for reproductive success.

My own take on complex systems—social systems, ecosystems, species—is considerably different. In contrast to the reductive, reproductively competitive ultra-Darwinian stance, I join other naturalists in seeing such complex systems as existing in their own right. For example, I agree with systems ecologists who understand ecosystems as complex entities. They believe ecosystems are formed of populations of many distinct species, held together by the complex flow of energy between different populations and the nonbiological environment. Ecosystems exist and can be studied in their own right, and it is mere empty rhetoric to claim that ecosystems ultimately are epiphenomena of competition for reproductive success.

Naturalists take the existence of large-scale systems seriously. We are loathe to write them off simply as epiphenomena of reproductive competition—especially entities like ecosystems, which lie squarely in the realm of economics (matter-energy transfer processes) and have nothing overtly to do with reproductive biology in the first place.

That's one major bone of contention, one way in which the ultra-Darwinians reveal a thoroughgoing reductionist stance. Ultra-Darwinians simply wave at large-scale systems, but only address the dynamics of gene-frequency changes as they see them, arising from competitive reproductive struggle, in what I call the "reproductive imperative."

Naturalists, in contrast, are attuned to the hierarchical structure of biological systems. They are convinced that there are processes relevant to understanding evolution that go on within each of these levels—from genes, right on up through populations, species, and ecosystems.

I have already touched on a related aspect of ultra-Darwinian reductionism: the credo that extrapolation of generation-by-generation change under natural selection is all we need consider when framing a theory of adaptive evolutionary change. But ultra-Darwinians are strangely silent about why adaptive change occurs when it does and why—for the most part during the history of a given species—adaptive change does not seem to occur. Nor has the problem here been ignorance of paleontological evolutionary patterns. I will give examples from both theories (especially the work of Sewall Wright) and patterns of distribution of living species that long ago should have shown evolutionary theorists that simple extrapolation of natural selection is insufficient.

The issue here is *context* of adaptive change. Naturalists such as myself completely agree that natural selection is the sole deterministic molder of adaptive evolutionary change (and is an important ingredient, as well, of adaptive stability). It is our contention, however, that evolutionary adaptive stasis and change occurs under sets of rules and circumstances, and that these rules and circumstances can be specified with some degree of precision. Especially crucial here is the notion of *species*, which I take to be collectivities of organisms sharing a common fertilization system. Ultra-Darwinians have all but renounced the very existence of species, and in any case pay them no special heed. In contrast, naturalists like myself see the nature, structure, and mode of formation of species as utterly essential to understanding evolution—certainly including the context of adaptive stasis and change.

It boils down to this: ultra-Darwinians emphasize continuity through natural selection and the primacy of active competition for reproductive success as the prime mover underlying absolutely all evolutionary phenomena. Naturalists, in contrast, see the complex biotic world as composed of discrete entities. Discontinuity is as important as continuity in depicting the real, natural world.

One final distinction between naturalists and ultra-Darwinians comes to mind—one which is perhaps less hard-and-fast than the others, but nonetheless worth noting. When looking for causes of phenomena—be they the generation of new species or the extinction of old ones—naturalists around the High Table will look to the physical, nonbiotic environment for the impetus nine times out of ten. We look to global cooling or meteor impacts to explain mass extinctions of the geological past. Ultra-Darwinians, in contrast, look to more purely biological causes for such phenomena. In this, the ultra-Darwinians are the conservatives. Darwin himself saw competition between organisms, and between species, as the driving force in much of evolutionary change, and most if not all extinction phenomena.

So who, then, are the ultra-Darwinians? They tend to be geneticists, absorbed with population-level phenomena. They are numerically more abundant than naturalists. The vast majority of the Society for the Study of Evolution would identify more with ultra-Darwinian credos than with naturalist postulates as I lay them out and examine them. This in itself comes as a surprise, not to those of us within the field of evolutionary biology broadly conceived, but to many interested parties outside the profession.

Among the ultra-Darwinians, the names that will keep cropping up in this narrative include the aforementioned John Maynard Smith, now in very active retirement, a student of the renowned polymath J. B. S. Haldane. Maynard Smith has made a number of key contributions to the discourse. Though I naturally focus on the disagreements we naturalists have with the ultra-Darwinians, I do openly acknowledge that population genetics in its modern ultra-Darwinian incarnation can be, and often is, extremely powerful *when restricted to its proper domain*, this is to say on the genetics of subgroups of species (populations) *especially* when competition for reproductive success is the phenomenon under study. That's the proper context for such work, by definition. It is only when ultra-Darwinians uncritically extrapolate to larger, or inappropriate, realms that their formulations begin to miss the mark.

Oxford University's Richard Dawkins, an Englishman like Maynard Smith, has already appeared, cast as the ultimate ultra-Darwinian. Dawkins's often biting prose and his penchant for the apt metaphor have made him the most visible and perhaps most widely known exponent of the ultra-Darwinian approach.

The American George Williams, recently retired from the Ecology and Evolution Department of the State University of New York at Stony Brook (but still very much active and in the thick of things), is my candidate as prime mover of ultra-Darwinism. As we have sparred over the years, I have found as much to be admired in Williams's thoughtful contemplation of the biological landscape as I have found him to be, at times, frustratingly wrong-headed.

These three, as beacons, symbols, even icons, exemplify and distill the essence of ultra-Darwinism. Others of this ilk will appear, and we should never forget the legions of ultra-Darwinian contributors and sympathizers as we keep our gaze focused across the Table, primarily at this particular threesome.

Of the naturalists, my own voice will be heard the most. The danger in delineating opposing camps is downplaying the disagreement within each group. Still, the game here is to characterize opposing positions, and to make the best possible case for the naturalist perspective, while striving to characterize ultra-Darwinism as accurately as possible. When faced with disagreements within either camp, I will sometimes mention it, but sometimes not. Insofar as the naturalist camp is concerned, this means that I will be speaking my own mind. This should not necessarily be construed as the opinions of others I have identified as "naturalists."

And who are these others? Stephen Jay Gould, of course, is one. He, like Dawkins for his side, has led the charge in bringing these issues to a more general audience, which is not privy to the day-to-day dialogue at the High Table. A fellow Columbia University (Steve was two years ahead of me in graduate school) and American Museum of Natural History–trained paleontologist, he has spent his entire career at Harvard. He has been an original thinker on evolutionary issues ever since I have known him. Although we like privately to disagree on any number of things, we have been closely associated in developing this distinctive naturalistic perspective ever since we teamed up to publish the first "punctuated equilibria" paper in 1972.

Elisabeth S. Vrba, also a paleontologist, was trained and initially employed in South Africa. Since the mid-1980s, however, she has been in the Department of Geology and Geophysics at Yale. Hers has been one of the most original voices in the naturalist camp over the past decade and a half, particularly her notions of the "effect hypothesis" (named, ironically enough, with George Williams in mind!) and "turnover pulse" of the interrelation between environmental change, migration of species, extinction, and true evolutionary change.

And, once again, I reiterate that those mentioned are merely the tip of the iceberg. All are important players in the game, to be sure, but by no means are they the only members of the naturalist camp to be seated at the High Table.

The Roots of the Matter: How Did the Evolutionary High Table Evolve?

Evolutionary biologists are fond of tracing their roots back to Charles Darwin—especially back to the pages of his epochal *On the Origin of Species*. Darwin's major task in the *Origin* was to convince his contemporaries that life has had an incredibly long history, and that all organisms on the face of the earth are descended from a common ancestor in the remote ages of the geological past. Although others before him had toyed with the notion of evolution, most of the best minds of the time still maintained the traditional consensus view that evolution had not, in fact, occurred.

The influential British philosopher-historian William Whewell reflected the consensus most graphically in his *History of the Inductive Sciences* (1837) when he wrote that "species have a real existence in nature, and a transition of one to another does not exist." Whewell was simply restating a famous maxim of Linnaeus (Carl von Linné), mid–eighteenth-century Swedish founding father of modern systematics (the science of biological diversity): "There are as many species as originally fashioned by the Creator ('Infinite Being')."

To establish the very idea of evolution, Darwin had to establish the possibility of connections linking those seemingly discrete entities that Linnaeus called "species." And therein lies a huge irony in the very title of Darwin's magnum opus. Darwin essentially destroyed the notion of species as real, discrete entities in nature. He had to, simply to establish the very fact of evolution, of the historical interconnectedness of all life.

To Darwin, species became simply progress reports in the history of life. At any given time, species might seem quite discrete, one from another. But the different kinds of birds in your backyard were not so different in past geological epochs; and, more than likely, through the ongoing evolutionary process, these species will become even more divergent in the geological future.

The essence of Darwin's vision was that stability is an illusion of the moment. As time goes on, as environments inevitably change, natural selection just as inevitably molds the heritable variation within a species, matching organisms' features to the altered environmental circumstances. Darwin believed evolution to be inevitable given the passage of time—so long as there is heritable variation. Even if environments remain relatively stable, natural selection will constantly be working to improve adaptations, honing a progressively better fit between organism and environment.

The Darwinian view sees change as inevitable. Species cannot help but change with the times. Far from remaining static, species are in constant flux, in effect evolving themselves out of existence. Slowly but inevitably species are altered. And when enough change accrues within a species lineage, we can say that a new species has evolved from its ancestor. In this model, species are just way stations along a continuous stream of modification.

Darwin began in a world that pronounced stability real and change illusory. He left just the opposite as his legacy: stability and discreteness had become the illusions, change and continuity the reality.

Given the worldview he was determined to replace, Darwin's argument makes abundant sense. And it was based on sound principles—including a healthy dose of empirical evidence. Darwin began his *Origin* with a discussion of his experiences as an apprentice pigeon breeder. He was convinced that animal husbandry and selective plant breeding had something important to tell us about how features of organisms could be modified in the natural world. By choosing only a subset of a generation—those with the desirable traits most in evidence— allowing them to breed, then repeating the process, breeders could change the attributes of bloodlines fairly quickly.

Darwin saw that artificial, domestic breeding in some ways mimics natural selection in the wild. Thus his basic syllogism: the modest changes that can be coaxed, by breeder and nature alike, from one generation to the next can be magnified a hundredfold over truly long segments of time.

It is no accident that Darwin, perhaps more than any other mid–nineteenth-century scientist, did more to establish the true order of magnitude of geological time. He counted earth history in the hundreds of millions of years, sometimes even overestimating the true antiquity of geological formations as later determined through isotopic dating. Darwin simply needed prodigious amounts of geological time to allow the gentle but persistent motor of evolutionary change—natural selection—to run sufficiently long to produce the panoply of living diversity on earth.

Darwin, of course, had his critics from the outset. Some focused on natural selection itself. It is often said that Darwin succeeded where others before him (such as his grandfather Erasmus) had failed precisely because he had, in natural selection, a credible mechanism to account for evolutionary change. Even so, it is nonetheless true that many of his early critics accepted the basic idea that life had evolved, and yet had a hard time accepting natural selection as its sole cause.

Whatever his early critics had to say, Darwin demonstrated without a shadow of doubt that natural selection is an ineluctable law of nature. Given heritable variation, and the fact that population sizes are limited, so all organisms of each generation cannot possibly manage to reproduce, it must be true that the transmission of traits from one generation to the next is inherently biased—in other words, selective. In general, what proved to work best for the parental generation will disproportionately be handed down to the next.

No naturalist evolutionary biologist seated at the High Table entertains the slightest doubt that natural selection is the deterministic process underlying adaptive evolutionary change. But how the process actually works in biological nature is a bone of hot contention between ultra-Darwinians and naturalist evolutionary biologists.

We naturalists can trace part of our intellectual heritage back to a few of the early commentators who felt that Darwin went too far in emphasizing continuity in his basic evolutionary imagery. "You have loaded yourself with an unnecessary difficulty in adopting *Natura non facit saltum* (nature never makes leaps) so unreservedly," wrote T. H. Huxley—Darwin's famous "bulldog"—to Darwin the day before publication of the *Origin*.

German biologist Moritz Wagner, writing a decade after the *Origin* appeared, was among the first to stress the importance of *isolation* in evolution. Zoologists knew that closely related species tend to replace one another geographically—meaning simply that the closest relative to any given species is often found some other place, living in a similar fashion, rather than in the same region. Eastern and western meadowlarks, two closely related species of North American birds that are seldom found living in close proximity, are a case in point.

Zoologists also knew that closely related species typically do not interbreed, no matter how closely they might resemble one another. These observations led British zoologist G. J. Romanes, writing in the late nineteenth century, to remark that "without isolation or the prevention of interbreeding, organic evolution is in no case possible." Romanes simply saw that the world consists of many different species, and that the evolutionary development of differences between species is correlated with the physical isolation of their ancestors.

But the role played by geographic and reproductive isolation remained a subdued theme in evolutionary biology for the first several decades of the twentieth century. Something more crucial was at stake. With the thrice-simultaneous rediscovery of the early genetic work of the Austrian monk Gregor Mendel in 1900, the fledgling science of genetics was quickly taken as a fundamental challenge to the very essence of Darwinism.

Darwin had been but one of a long series of biologists who could make no real headway in understanding heredity. Indeed, Darwin's own ideas on the subject were far off the mark. He thought that hereditary factors were drawn from all parts of the body, in particles he called "gemmules," to be concentrated in eggs and sperm. German biologist Auguste Weismann had, by the late nineteenth century, disproved such Darwinian *pangenesis* when he showed that whatever the controlling factors of inheritance might be, they are confined to what he called the "germ" tissues—ovaries and testes, for example, in vertebrate animals.

The facts that organisms vary one from another, and that organisms tend to resemble their parents, were two observations absolutely vital to Darwin's concept of natural selection. That he had absolutely no idea—or rather, that he had the wrong idea—about how biological inheritance works and yet managed to formulate the concept of natural selection, is crucially significant to an understanding of how biological systems are organized, and what underlies many of the arguments in modern evolutionary biology. How can an idea so dependent on the fact of biological heredity possibly be correct when the basis of heredity was so completely misunderstood?

Darwin, in effect, treated the inner workings of heredity as a black box. All he needed to support his idea of natural selection was the realization that there is heritable variation within species. And though he himself quite naturally wished to understand the biological basis of heredity, his desire to do so actually says nothing of the importance of such knowledge to the formulation of the concept of natural selection. That he got it wrong merely underscores the independence between the mechanics of heredity and the process of natural selection.

Yet when Mendelism was rediscovered and the fledgling field of genetics burst into prominence in the first decade of the twentieth century, many of the early geneticists felt that Darwinism had been superseded. For one thing, biology now seemed more "scientific." Genetics was just a part of an exciting new wave of discoveries in physiology and biochemistry achieved through direct experimentation in the newly electrified laboratories. Biologists of the day openly crowed that they had finally gotten past the amateurish scientific dabbling known—disparagingly by then—as "natural history."

But there was also the belief that a more direct grasp of how the process of inheritance really works would automatically supersede earlier theories of the evolutionary process. It seemed to many of the earliest geneticists that a more precise understanding of the genetic material—what it is, where it is located, how it is transmitted, and, critically, how it can change—would inevitably lead to far more precise explanations than the more qualitative and vague notion of natural selection.

Early geneticists quickly stumbled on the process of mutation, a change in heritable features that pops up suddenly and spontaneously. Dutch botanist Hugo DeVries, working with the evening primrose (*Oenothera*, a North American–derived weed), noticed occasional large-scale changes in the color and structure of isolated flowers on otherwise normal plants. All it would take for a new species to arise, DeVries thought, would be for some comparable mutation to arise spontaneously, as seemed to be happening in his evening primroses.

If mutation could explain the origin of heritable change, why be encumbered with the excess baggage of natural selection? Some early geneticists really felt they had a superior theory of evolutionary change staring at them with their newfound understanding of the mechanisms of heredity and the origin of new genetic variation through mutation.

The early geneticists saw a number of conflicts between the Darwinian vision and the newfound facts of genetics. Darwin spoke of gradual changes in evolution with natural selection working on a continuous array of variation. Average height within a species, for example, might be expected to increase gradually, given a spectrum of variation in individual heights. This is analogous to the gradual increase in milk yield that a patient animal husbandryman could elicit in a breed of cattle by letting only those cows who produce the highest volumes of milk reproduce. But, starting with Gregor Mendel, the early geneticists saw inheritance as basically particulate, coming in sharply defined alternate states, such as red and white flowers on a single plant. Even if the alternate forms of genes (now called alleles) producing red and white flowers could be combined to form pink ones, when plants with pink flowers reproduce, the red and white factors would be sorted out and passed along separately to the next generation—which would be 25 percent red, 25 percent white, and 50 percent pink. No continuous spectrum of variation there.

Moreover, Darwin had anticipated slight, heritable variations appearing—variations that might prove beneficial to organisms and thus targeted by natural selection. In contrast, early data from the geneticists seemed to show that mutations had large-scale effects, and were often harmful, even downright lethal.

No doubt about it, the first two decades of the twentieth century were rough on Darwinians. Confusion was so rampant that many biologists—including paleontologists like Henry Fairfield Osborn, President of the American Museum of Natural History—felt free to write their own evolutionary theories, with blatant disregard for the real progress that had been made in genetics. Osborn simply invented his own theory ("Aristogenesis") to fit his idiosyncratic view of evolution. The essence of his notion was that forces internal to organisms yield a relentless drive for improvement during evolutionary history.

Restoration of the core Darwinian view was the great achievement of Ronald Fisher, J. B. S. Haldane, and Sewall Wright, the founders of neo-Darwinism. By the 1920s, it had finally become clear that some mutations are indeed small-scale in effect, and that not all mutations are necessarily harmful. The hereditary basis of many continuously varying traits—such as height or milk yield—have much more complicated genetic bases than the simple dominant-recessive picture in the earliest genetic models. Largely through their mathematical skills, Fisher, Haldane, and Wright were able to reconcile the Darwinian notion of evolution by natural selection with the mechanisms of heredity as they were known circa 1930.

Natural selection once again was firmly established in the minds of virtually all serious evolutionary biologists as the central molder and shaper of adaptive, evolutionary change. Reconciliation of the particulate, discontinuous nature of hereditary particles—genes—with the continuous spectrum of adaptive variation seen in the physical properties of organisms themselves reestablished the Darwinian preference for continuity and revealed the importance of discrete levels of biological phenomena.

For natural selection, in a fundamental sense, doesn't care how heredity works: how genes are faithfully copied, how and why mutations arise, or what the molecular basis underlying a particular gene product might be. That Darwin was able to grasp and formulate so fundamental a truth as natural selection without having the slightest idea what causes organisms to resemble their parents, or why new heritable variation can appear, starkly sets off the existence of two related, yet independent realms. One is the genetics of individual organisms, where cells divide, their genetic components are duplicated and translated, and mutations arise. The other is the genetics of populations, where frequencies of different forms of genes (alleles) change from generation to generation through essentially chance factors (so-called genetic drift—a concept first developed by American geneticist Sewall Wright in the 1930s) and through natural selection.

The great Russian-born biologist Theodosius Dobzhansky made just that point in 1937. I think of Dobzhansky as perhaps the last great swing figure in evolutionary biology. He started out as a naturalist, a specialist in ladybird beetles, but left his native Russia to come to Thomas Hunt Morgan's labs at Columbia to study genetics. It was the Morgan group that had made the lion's share of the big discoveries in the first two decades of genetics. It was there, for example, that genes were identified as particles ensconced on chromosomes housed in the nuclei of plant and animal cells. By the time Dobzhansky got to Columbia in 1927, the heady pioneering days were over.

Unlike virtually all his predecessors in genetics, Dobzhansky refused to draw evolutionary conclusions from his laboratory experiments on fruit flies unless he could verify his results in wild populations of fruit flies out in the field. His monumental series of 43 papers on *The Genetics of Natural Populations*, begun in 1937 and spanning some 38 years, were the direct product of his convictions.

In 1937, Dobzhansky published *Genetics and the Origin of Species*, a masterful application of the newly forged ideas of neo-Darwinism to evolutionary patterns in the real world beyond the confines of the lab. Dobzhansky was a vital innovator in the world of evolutionary genetics, but was, unlike the vast majority of his contemporaries and successors in the genre, as much at home in the natural world as the lab. Indeed, Dobzhansky is as much a patriarchal figure to our naturalist perspective as he is to population geneticists and today's ultra-Darwinians.

Dobzhansky made much of the seeming independence—or "decoupling"—of natural selection from any system of precise formulations of the principles of heredity. To Dobzhansky, the distinction was both clear and easy to understand: genes are replicated and translated within individual organisms and transmitted, in the case of sexually reproducing organisms, in the mating process. Natural selection, on the other hand, takes place in the context of entire populations of males and females of the same species. He called the hereditary process proper "physiological genetics" and the context of natural selection is, accordingly, "population genetics."

Individuals are parts of populations, and populations are parts of species. Dobzhansky was simply saying that biological nature is hierarchically organized into distinct levels. Within each level there are certain "rules"—certain kinds of processes intrinsic to that level that are different from those rules and processes at other levels. That's why explication of the process of natural selection does not need to include an explanation of the rules of inheritance. The two sets of processes, both vital to the evolutionary process, take place at different levels.

Dobzhansky's explicit acknowledgment of the hierarchical structure of biological systems was a great step forward. Recognition of the crucial difference between genetic processes within individuals versus those taking place within populations underlay the successes of Fisher, Haldane, and Wright in reconciling the Darwinian vision of evolution through natural selection. It is a distinction that all evolutionary biologists—naturalists and ultra-Darwinians alike—openly embrace.

But Dobzhansky went one step farther. He saw that populations are actually parts of still larger systems—specifically, species. In so doing, he resurrected the long-neglected notions of geographic and reproductive isolation, and, in so doing, introduced an aspect of *discontinuity* that had remained muted ever since the earliest days of evolutionary biology.

Dobzhansky argued that Darwin in effect was wrong to have minimized the discontinuity in physical features of organisms. Most critically, he was wrong not to have emphasized the general inability of interbreeding between closely related species. Darwin never dwelt on evolutionary branching, the proliferation of species, in any great detail. He imagined that closely related species, living at the same time, had drifted apart, continuing to accrue differences so that interbreeding finally ceased between them.

Dobzhansky felt that the inability of two closely related species to interbreed is a direct product of the evolutionary process. There is a reason, beyond sheer accident, why species evolve into discrete reproductive communities. To Dobzhansky, fragmentation of an ancestral species into two (or more) discrete reproductive communities enabled each fledgling species to eliminate variation, to focus more closely on specific adaptive requirements of the environments in which they lived.

We'll pick up these themes in later chapters. There are many, including Dobzhansky's contemporary Ernst Mayr, the great contributor to the theory of geographic speciation, who have taken sharp issue with the details of Dobzhansky's characterization of the speciation process. Dobzhansky stressed that isolation between species is a direct outcome, an actual result, of the evolutionary process itself. Discontinuity is as much an evolutionary phenomenon as is the more familiar, traditionally stressed continuity that Darwin worked so hard to establish.

Ornithologist Ernst Mayr quickly followed Dobzhansky with his *Systematics and the Origin of Species* (1942). Mayr, like Dobzhansky, is a European immigrant steeped in the traditions of continental biology. He was keenly aware of the early work of Wagner and of such later German biologists as E. Streseman who also stressed the importance of isolation in the evolution of birds. Mayr pointed out that Darwin's title was a misnomer. Nowhere in the *Origin* did Darwin actually discuss the derivation of one species from another. Rather, his picture of slow transformation suggested that species are not discrete entities after all, but merely stages along a continuous stream of evolutionary change.

That would not do for Mayr, any more than it satisfied Dobzhansky. Mayr likened species to individual *Paramecium*. Looking down the barrel of a microscope at a collection of *Paramecium* individuals is, Mayr declared, very much like contemplating species in nature. To be sure, one encounters individual paramecia in the reproductive process of dividing in two. Finding one in the middle of division, Mayr said, would make the question, How many paramecia individuals do I see? Difficult to answer. But the overall picture is simple enough, according to Mayr. You start out with a single individual, and, when the process of division is complete, there are two individuals where once there had been one.

And so it is with species. To the question, Are species real entities in nature? Mayr's response was unequivocal: if species aren't real, why then have a theory to explain their origin? With this, discreteness at the level of species had, finally, reentered scientific thinking for the first time since Whewell had pronounced species as having "a real existence in nature." Darwin needed to destroy the notion of species as real, discrete entities in order to establish the truth of his larger proposition: that life has evolved, that all organisms on earth are descended from a common ancestor. It took Mayr and Dobzhansky to affirm that Darwin's views on natural selection were correct, but that the commonsense observation that species are discrete entities—the hallmark observation of the pre-Darwinian era—was also correct.

In the late 1930s and early 1940s, paleontologist George Gaylord Simpson also entertained strong views on discontinuity in evolutionary history. Simpson was a curatorial colleague of Mayr's at the American Museum of Natural History in New York—a mere 40 city blocks or so south of Schermerhorn Hall on the Columbia campus, home of Morgan's famous "fly room" and home, as well, to Theodosius Dobzhansky.

Simpson saw nothing particularly wrong about Darwin's views on the gradual derivation of one species from another. He also went right along with Darwin's reasoning about why so few examples of gradual transformation are actually encountered in the fossil record: it is the fault of the record itself. After all, Darwin mused, the chances must be vanishingly small that any single organism will, upon death, escape the ravages of consumption, decay, and decomposition to become buried, its hard parts faithfully preserved for millions of years (escaping chemical dissolution or outright destruction in metamorphism), only to be exposed eons later, collected (prior to eroding away to bits) by an intelligent observer who sees to it that the fossil ends up on a competent paleontologist's bench!

But Simpson could blame only so much on the fossil record for gaps in the evolutionary histories of many different groups of organisms. For example, the earliest known bats and whales show up in rocks of the Eocene Epoch, roughly 55 million years old. Each, though certainly primitive forms of bats and whales, were nonetheless obviously true members of their groups. Not only could those Eocene bats fly, but they had already evolved the complex inner ear structures used by modern bats for echolocation.

There are no known intermediates between bats and their nearest land-dwelling relatives, and we are only beginning to find early whales that could still locomote on dry land. Fossils intermediate between two related but adaptively divergent groups tend to be as rare as hens' teeth. But, Simpson argued, these gaps are real, and they tell us something about the evolutionary process itself. How so?

Fifty-five million years is a long time, and both the whale and bat lineages diversified and, along the way, accumulated a measure of anatomical change. But that measure of change translates to a fairly slow overall rate of transformation over 55 million years. Here's Simpson's clincher: if we take that rather modest rate of bat or whale evolution over the past 55 million years and extrapolate it backwards to estimate when either group diverged from four-legged, fully terrestrial mammalian ancestors, we get an absurd result. If those evolutionary transitions occurred at the same rate as subsequent observed bat and whale evolution, it would have taken far longer than 55 million years for the first bats or first whales to have evolved from terrestrial ancestors—simply because the degree of anatomical transformation is so very much greater. But that, Simpson pointed out, is simply impossible. Bats and whales are placental mammals, and placental mammals evolved no earlier than 100 million years ago. There is no escaping the conclusion that the transition from primitive terrestrial mammals to the earliest bats and whales must have occurred at a much faster rate than subsequent whale and bat evolution.

Simple extrapolation of conventional evolutionary rates, in other words, just doesn't work to explain such patterns. Simpson claimed that the relatively abrupt appearances of large-scale groups—higher taxa, such as bats (Order Chiroptera) and whales (Order Cetacea)—are the rule rather than the exception. His theory to explain this pattern of abrupt and rapid evolution—*quantum evolution*—was consistent with contemporary genetic theory, but was formulated in direct response to a pattern of apparent discontinuity seen over and over again in the fossil record.

Simpson's initial formulation of quantum evolution in 1944 was bold and imaginative. He drew on Sewall Wright's concept of genetic drift—essentially a chance mechanism that alters gene frequencies from generation to generation without the guiding determinism of natural selection. Choosing the abrupt appearance of grass-eating horses in the Miocene epoch, Simpson theorized that a small population of ancestral, leaf-browsing horses lost their adaptive focus on leaves, and possessed variation in the direction of higher, broader many-cusped teeth. This was the requisite dental machinery for eating the newly evolved tough siliceous grasses. Genetic drift, Simpson argued, would help shift the population away from leaf-browsing, and natural selection would take over, working rapidly to alter horse teeth to much greater grazing efficiency. It was all-or-none, sink-or-swim evolution—and it had to move rapidly if it were to work at all.

I have long found Simpson's style of evolutionary analysis inspirational. Until Simpson came along, if a paleontologist found that patterns of stability and change in the fossil record didn't jibe very well with the Darwinian canon, the conclusion was always the same: the record is to blame. Simpson taught us to take the fossil record more seriously as a direct reflection of evolutionary history. With appropriate caution, we should test our theories against the record, and not automatically conclude it is the record's fault when, time after time, predictions fail to be confirmed when we tackle pattern.

Simpson keenly felt the mutual mistrust between paleontologists and geneticists seated around the High Table of the late 1930s and early 1940s. In a trenchant passage that foretells Maynard Smith's words cited at this chapter's opening, Simpson wrote in 1944:

> Not long ago paleontologists felt that a geneticist was a person who shut himself in a room, pulled down the shades, watched small flies disporting themselves in bottles, and thought that he was studying nature. A pursuit so removed from the realities of life, they said, had no significance for the true biologist. On the other hand, the geneticists said that paleontology had no further contributions to make to biology, that its only point had been the completed demonstration of the truth of evolution, and that it was a subject too purely descriptive to merit the name "science." The paleontologist, they believed, is like a man who undertakes to study the principles of the internal combustion engine by standing on a street corner and watching the motor cars whiz by.

Plus ça change, plus c'est la même chose. But Simpson was right. We paleontologists can make contributions to understanding how the evolutionary process works by identifying recurrent patterns in evolutionary history—patterns that call for unique combinations of genetic processes otherwise undreamt of by geneticists restricted to small-scale experiments carried out over at most only a few years. As Simpson said, such experiments "may reveal what happens to a hundred rats in the course of ten years under fixed and simple conditions, but not what happened

to a billion rats in the course of ten million years under the fluc-
tuating conditions of earth history."

Fair enough. But Simpson went on to say: "Obviously the
latter problem is much more important"—an unfortunate bit of
genetics baiting that actually denigrates experimental verification of natural selection. Clearly both approaches are important.

Confluence and Consensus?

The years immediately following World War II ushered in an era
of the most remarkable consensus and calm in evolutionary biology since the late nineteenth century. George Simpson remarked
on the nearly insurmountable difficulties of writing a book during wartime in the preface of his *Tempo and Mode in Evolution*
(1944). (Simpson, like so many of his colleagues, served on active
duty, most notably as an intelligence officer under General
George Patton. Simpson reportedly refused to comply with
Patton's direct order to shave off his beard, on the grounds that
he only took orders from his immediate superior officers. It is
difficult to imagine Simpson admitting he had superiors of any
kind.)

In any case, as the flow of normal life resumed after the war,
there was much catching up to do. Everyone, it seems, read
everyone else's books: there were Mayr's (1942) and Simpson's
(1944) to confront. But there had been wartime developments in
Europe as well. Julian Huxley—grandson of Thomas Henry and
brother of Aldous—published *Evolution: The Modern Synthesis* in
1942. German ornithologist Bernhard Rensch's *Neuere Probleme
der Abstammungslehre* (ultimately translated into English as *Evolution Above the Species Level*) appeared in 1947. Rensch's central
theme was that there is nothing known about large-scale evolutionary phenomena, such as the development of whales from
four-legged terrestrial mammalian ancestors, that is inconsistent
with the basic neo-Darwinian theory of evolution through natural selection.

Rensch had articulated the general postwar credo: There is nothing in biological nature—in the data of the genetics of natural populations, in systematics, in paleontology—to suggest that there are any evolutionary processes other than natural selection working on the natural genetics of variation within populations. Steve Gould has dubbed this postwar movement the "hardening" of the synthesis. It was a narrowing of the relatively richer variety of concepts that had typified evolutionary biology in the 1930s.

Sewall Wright was probably the biggest loser in postwar population genetics. The trend, both within population genetics and among the more extended crowd of evolutionary biologists, was to endorse natural selection as the dominant force shaping evolutionary change. Wright's genetic drift was relegated to lip service treatment in virtually all quarters. Years later, when Ernst Mayr convened a large collection of biologists, historians, and philosophers to examine the roots of the "modern synthesis," Wright was conspicuous in his absence, because "everyone knows what Sewall has to say" (as Mayr was widely reported to have said).

Oddly enough, Wright—the consummate mathematical theoretician and analyst of experimental results never known for his direct experiences with nature—had much of lasting importance to say about the structure of species in nature. His name shall reappear at our side of the High Table as the story unfolds.

But Wright himself helped bring about the "hardening" of the synthesis. In 1945, writing in (of all places) the journal *Ecology*, Wright tore apart Simpson's arguments for quantum evolution in his review of Simpson's *Tempo and Mode in Evolution*. Ridiculing a mistake in mathematical notation, and disparaging Simpson's grasp of Wright's own theories, Wright summarily dismissed Simpson's attempt to reconcile contemporary evolutionary genetics theory with patterns in the fossil record. So much for Simpson's passionate conviction that patterns in the fossil record are phenomena begging to be addressed by evolutionary theorists.

And so much for détente. Wright's critique was especially important because Simpson had utilized Wright's notion of genetic drift as part of his explanation of how major adaptive shifts—leading to the appearance of wholly novel groups, such as grass-grazing horses—can happen very rapidly. I have always thought that Simpson was stung by Wright's critique, and that this in part explains his famous retreat from his bold stance of 1944 to a far more conventional depiction of large-scale evolutionary change. By 1953, in his *The Major Features of Evolution*, Simpson was perfectly willing to attribute all evolutionary phenomena—his "major features" of evolution—directly and solely to the action of natural selection, with none of the ifs, ands, or buts of his earlier work. Simpson's capitulation is the most difficult aspect of the hardening of the synthesis for me, a naturalist paleontologist, to accept.

Why this rush towards consensus? For one thing, no matter how disparate the particular views among the evolutionary biologists of the 1930s and 1940s, the overall intent of such figures as Dobzhansky, Mayr, and Simpson had always been to show that their data were indeed consonant with the basic neo-Darwinian principles worked out by Fisher, Haldane, and Wright. And neo-Darwinian population genetics itself stressed Darwin's original focus on selection as the central shaper of evolutionary change.

Then, too, there was a certain unity gained from common resistance to an eclectic group of dissenters. These dissenters were a melange of avowed or accused anti-Darwinian professional biologists who, for a variety of reasons, maintained the antiselectionist stance that had first appeared in Darwin's own time. The German paleontologist Otto Schindewolf, for example, stressed abrupt appearances of major new groups—the very same pattern that Simpson addressed in his own theory of quantum evolution. But Schindewolf insisted that such patterns of discontinuity were inherently non-Darwinian, and required theoretical explanation radically different from Darwinian natural selection.

Then there was Richard Goldschmidt—everyone's whipping boy. Goldschmidt, a wartime refugee from Hitler's Germany, was a sophisticated geneticist who had devoted much of his early career to the genetics of the silkworm moth, *Lymantria dispar*. Like Simpson and Schindewolf, Goldschmidt was struck by a pattern of discontinuity in his data—this time, involving genetic variation. The characteristics that separate closely related species of moths are not the same features that Goldschmidt saw varying *within* species. How could this be if the evolution of new species flows smoothly from natural selection working on variation within species?

Goldschmidt imagined that there must be large-scale mutations that account for the differences between species. These large-scale mutations could occasionally produce, rather abruptly, major new evolutionary groups. He even coined a catchy phrase—"hopeful monsters"—capturing the sense of his idea of rare, large-scale mutations that just might, every once in a great while, not prove lethal, but instead instantaneously create a new, successful evolutionary group. Goldschmidt, along with paleontologist Schindewolf, were "saltationists," people who see evolution proceeding largely by sudden "jumps."

Goldschmidt's ideas, quite naturally, were anathema to the newly emergent neo-Darwinians. After all, they had just buried the old DeVriesian idea that evolution proceeds mostly through mutation alone, along with a number of other purported inconsistencies between genetics and natural selection. In their critiques they stressed the lack of any empirical support for Goldschmidt's ideas, and "hopeful monsters" quickly became a term of derision. Goldschmidt was still being mocked at Columbia in the 1960s when I went through the program.

The specter of Goldschmidt's hopeful monsters even haunted us in the late 1970s, when it suited many biologists, including Ernst Mayr, to equate our notion of punctuated equilibria with Goldschmidtian saltationism. But punctuated equilibria (as we shall see), reduced to its barest essentials, was a simple attempt to address patterns of discontinuity between species in neo-Darwinian terms in a way derivative of, but differing in detail from, the work of Mayr and Dobzhansky, on the one hand, and Simpson on the other. Gould and I are no more Goldschmidtian than were Simpson, Mayr, and Dobzhansky.

But such was the galvanizing effect of dissent—with Goldschmidt's foremost—that the neo-Darwinian wagon train would readily pull into a tight circle and ward off the attacks of any and all who dared question the central premise that evolution is first and foremost a matter of natural selection honing adaptations within populations. Within that camp, the neo-Darwinists tended to claim that this is essentially all there is to evolution. The early naturalists—the field geneticists, systematists, and paleontologists, as well as some ecologists—were more prone to insist things were a bit more complicated than the single-level population geneticist view centered around Ronald Fisher. But, as time wore on, even the naturalists began to succumb. The mechanics of the evolutionary process per se, it came to be acknowledged by the early 1950s, really do lie wholly in the province of population genetics. All a paleontologist or systematist can really do is attest to the "fact of evolution"—as Simpson put it in 1944 as he prepared to take strong issue with that very sentiment.

The Society for the Study of Evolution was founded in 1946. A milestone conference was held at Princeton in 1947, during which geneticists, paleontologists, systematists, and other biologists got together and agreed, in effect, that the neo-Darwinian paradigm was both necessary and, in the main, sufficient to explain evolution. Though a number of paleontologists served as the society's president and contributed to its journal *Evolution*, the pages of that august publication rather quickly came to be filled with the contributions of evolutionary-minded population geneticists. Paleontologists still belong and still occasionally contribute. I published the forerunner announcement of what we later called "punctuated equilibria" there in 1971, and Steve Gould has even served as the society's president. But these are exceptions to the rule that paleontology and systematics have figured far less in the work of the society than originally envisioned by its founders in the late 1940s.

Through the 1950s, consensus grew stronger still. Only a few biologists, such as Chicago paleontologist Everett Olson, really challenged orthodoxy. Few saw fit to challenge the essentially reductive notion that all evolutionary phenomena of any import were not only compatible with, but were to be understood purely in the terms of, natural selection within populations.

But that is consensus only in the narrow sense. Gone was any real attention to patterns of discontinuity at levels above the population. Simpson abandoned the guts of quantum evolution when he redefined the process as essentially an "extreme and limiting case of phyletic change"—meaning linear evolution under the complete and direct control of natural selection. Even Dobzhansky, who had been the first to inject discontinuity into the discourse in the late 1930s, downplayed the theme in the third edition of *Genetics and the Origin of Species* (1941). Only Mayr was left to champion the importance of species in evolution in the 1950s, and his next big book on the subject was not to appear until 1963: *Animal Species and Evolution*.

As the 1959 Darwinian centennial approached, everyone began congratulating one another that Darwin had been vindicated, and that at last a comprehensive and essentially complete evolutionary theory was at hand. No one seemed to realize that the "consensus" was really little more than hegemony achieved by a Fisherian population genetics view of the evolutionary process. And no one foresaw the dichotomy soon to arise between emboldened Fisherians—who were to take population genetics to even deeper reductive depths—and latter-day naturalists— who quickly appeared to sing the song of discontinuity and to raise the importance of large-scale biological systems in the evolutionary process.

None, that is, save a very few, like paleontologist Everett Olson and philosopher Marjorie Grene. In 1958, in a paper explicitly marking the Darwinian centennial, she had the temerity to propose that Otto Schindewolf's theory was "more preferable" to Simpson's on philosophical grounds, because Schindewolf paid more attention to the empirical facts of the matter. She had been looking at Simpson's later work, in which he essentially recanted his earlier hard look at discontinuity, and understandably felt that Simpson was pushing a story of continuity in spite of abundant evidence to the contrary. Grene's paper gave me a headache as a student, coming to it full of expectation that her attack on the renowned Simpson would of course be patently flawed. Hard as I tried, I could detect few chinks in her armor, and I turned to some of the published rejoinders, including some penned by my teachers, only to find that they, too, offered no fully satisfactory defense of Simpson.

But Grene was virtually a voice crying in the wilderness. Evolutionary biology circa 1959 was little more than a statement that Fisher's conception of natural selection is the necessary and sufficient engine of evolution. Darwin, everyone was saying, had had it right all along.

The Table is Set and the Guests Arrive

Maynard Smith was very right when he suggested that paleontologists were pure arrivistes at the High Table in the early 1980s. Evolutionary theory had long since become the exclusive domain of neo-Fisherian population geneticists. Nor was the absence of paleontologists, systematists, ecologists, and other naturalists a matter purely of haughty exclusion. After all, if you buy the argument that the true mechanics of the evolutionary process lie squarely, and uniquely, within the province of genetics, and if you happen to be a paleontologist, you will feel uncomfortable—perhaps even unwelcome—at the High Table.

But, if you feel that there are patterns in nature—for example, patterns of discontinuity between species or between higher taxa—you might feel (as Simpson, Dobzhansky, and Mayr originally did) that the whole story cannot possibly lie in the realm of the genetics of populations. The grand mistake, the cardinal sin that rightly carries automatic suspension of seating privileges at the High Table, is to suggest a theoretical proposition that assumes that the neo-Darwinian paradigm is somehow erroneous. Theories that claim, in some fundamental sense, to be alternatives to the neo-Darwinian paradigm bear an immense (and I believe insurmountable) burden of proof on their metaphorical shoulders.

Ultra-Darwinians who people the following chapters are fond of accusing their opponents of precisely that crime: opposing the core neo-Darwinian paradigm. That's precisely what Henry Fairfield Osborn did with his theory of "aristogenesis" in the 1920s, and that, too, is the original sin of Richard Goldschmidt. But that's not what we latter day arrivistes at the Table are doing. We are making the milder, far more defensible, claim that one cannot simply take the neo-Darwinian paradigm and extrapolate it all across the board. We simply do not believe in attributing population-level phenomena to such disparate entities as species, higher taxa, social systems, and ecosystems. And that's precisely what the ultra-Darwinians have been up to, claiming all the while that they've actually demonstrated the truth of what was in reality assumed all along: that the neo-Darwinian paradigm of the workings of natural selection within populations is necessary and sufficient to explain evolutionary history. We naturalists agree that it is necessary. But it isn't sufficient.

Naturalists argue that ultra-Darwinians look endlessly at the minutia of population biology, periodically glancing up to apply their principles across the board in simplistic and generally unrealistic ways. Never shy, ultra-Darwinians persist in telling anyone who will listen that our naturalists' attempts to formulate theoretical propositions on the nature of the evolutionary process are for the most part overwrought failures.

It sounds like George Simpson's story of geneticists versus paleontologists all over again. But there has been progress, within both the ultra-Darwinian and modern versions of the naturalist camps. The problem is reconciling the two. Anyone not seated at (or in the immediate vicinity of) the High Table would surely be reminded of the blind men and the elephant. And while the spirit of the present enterprise remains argumentative, I share George Simpson's conviction (and Maynard Smith's sometime view) that surely someday, somehow a genuine rapprochement will emerge. One day evolutionists will be able to dine in relative harmony at the High Table.

2

The Heart of the Matter

Adaptation and Natural Selection

Adaptation is the very heart and soul of evolution. It is *the* scientific account of why the living world comes in so many shapes and sizes: how the giraffe got its long neck, why porpoises look so much like sharks and the extinct ichthyosaurs, how birds fly, and literally millions of similar questions. The only other account of this spectacular display of diversity is the creationist tale: that a supernatural Creator fashioned the world, including its organic contents, the way we find it. But that form of explanation, by its very nature, lies outside the bounds of the scientific enterprise.

Some organisms match their surroundings so intimately that it can be hard to imagine how the evolutionary process could have fashioned such a perfect match. The multicolored leafy sea dragon (a species of "seahorse" fish) looks just like a chunk of the algae ("seaweed") in which it hides. The complexity of the human eye was the rallying point for Darwin's earliest severe challenge. How can such intricate structures evolve through a series of less perfect, less fully functional intermediate stages? So goes the plaint, which for the most part is now abandoned in evolutionary circles, but is still a prominent feature of creationist tracts.

"Natural" selection is Darwin's term, counterpoised with "artificial" selection. Design, according to pre-Darwinian, religiously imbued tradition, implies a Designer. But, countered Darwin, breeders in a very real sense are themselves designers, drawing forth desired attributes while dampening others by allowing only a select few to breed. (Biotechnology today just ups the ante, intensifying the selectivity and, in some instances, eliminating the need to wait for generations for the desired effect to occur). Nature, in effect, does the same thing—and there is a sense of natural selection in this active mode to be found in Darwin's *Origin of Species*.

On the other hand, the basic, original sense of natural selection does not in fact see Nature as the natural analogue of the human animal breeder. More organisms are produced each generation than possibly can survive and reproduce. The qualities that convey relative success in simply surviving—in making a living and avoiding being eaten, ravaged by disease, or victimized by climate—tend to be passed along to the next generation. Reproductive success is a by-product of economic success, and natural selection is a passive filter of which heritable features get passed along to the next generation, simply because not all members of the previous generation will survive and reproduce. Though Darwin occasionally presented natural selection in the active voice, as a Natural Artificer, for the most part he saw natural selection as a fallout, a simple reflection of "what worked better than what" in the preceding generation.

And so, goes the core canon, if the environment remains the same, the adaptations of organisms—their behaviors, physiologies, and physical features—will be honed to fit their surroundings even better than they did before. If the environment should change, then natural selection will track that change and adaptations will change—provided, of course, that there is sufficient genetic variation available in the population.

Now, I disagree with the ultra-Darwinians on many things, but we all agree on the core fact of adaptation. That is, the basic way that natural selection acts to shape, hone, and conserve adaptations—though conservation of adaptation is a point wholly ignored until recently by ultra-Darwinians.

But there is something to the old cry that evolutionary biology has paid little attention to intermediate stages—whether in the evolution of the eye or all the stages that sea horse lineage had to go through to look so much like the foliage it lives in. Anatomists, impressed with the complexity and stability of the structures they study, are notoriously unable to imagine how those structures could change over time. Yet all evolutionists know that these features did change: vertebrates are allied with starfish and a number of other invertebrate groups—none of which have heads, let alone eyes. The vertebrate eye most certainly did evolve.

But, on the other hand, all we have in traditional evolutionary biology is this extrapolation of the basic principles of natural selection over time to account for all manner of adaptive change. Ultra-Darwinians simply have not considered the *context* of adaptive change to any great or effective degree. This is the subject of the following three chapters.

In the remainder of this chapter, I will review three arenas in which the goal has been to sharpen the actual study of adaptation. In the first, the ultra-Darwinians set out to place the study of adaptation on a more firm scientific foundation, but end up, to my naturalist's eyes, distorting the basic original meaning of natural selection. I then turn to two other approaches to adaptation, one aimed directly at the ultra-Darwinians, the other coming from a third sector, which professes not to care about adaptation at all!

Ultra-Darwinism, Natural Selection, and Adaptation

Ultra-Darwinians tend to see themselves as archdefenders of Darwinian tradition. George Williams wrote his *Adaptation and Natural Selection* (1966), the cornerstone articulation of ultra-Darwinism, with the avowed purpose of countering the suggestions of Australian biologist V. C. Wynne-Edwards. Wynne-Edwards had questioned a cornerstone assumption of natural selection and proposed instead a form of "group selection" to explain certain aspects of adaptation. In essence, Wynne-Edwards proposed that selection might act for the good of the group and not necessarily for the good of the individual. Williams replied with a vigorous attack, maintaining that all aspects of organismic design could be explained satisfactorily through natural selection. There would be no need to invoke more complex, difficult to establish, and conceptually muddled notions of "group" selection.

In an evolutionary sense, in other words, Williams maintained that everything can be explained in terms of the "good of the individual." He saw no need to construct arguments about the "good of the group," especially the "good of the species." Selection, Williams felt, has no eyes for the future—no way of telling what might be advantageous for the continued survival of a species. Selection is based solely on the heritable properties of organisms, and records only what worked best for organisms—not the entire population or species—in preceding generations. I'll return to aspects of the group selection debate in chapter 5.

But it is by no means true that ultra-Darwinians have done nothing to alter the essential Darwinian canon of adaptation through natural selection. On the contrary, natural selection has been reformulated in a subtle, yet profound way to emerge as an active process to a degree unimagined by Darwin. In the past 30 years or so, ultra-Darwinians have reformulated their basic conception of natural selection, changing it from a passive to a directly active agent operating in the natural world, pushing the evolutionary process relentlessly forward.

The contrast between Darwin's and, say, George Williams's view on natural selection is fairly stark. Once again, Darwin saw that in a world of finite resources and real perils, only a subset of a generation will manage to survive and reproduce. On the whole it will be those individuals best able to meet life's exigencies who will be doing the reproducing—factoring out the random component of luck, of course. The standard example pictures a baleen whale cruising, mouth open, through a krill population. Luck, not adaptation, determines which krill die and which live on. But over and above pure chance, the heritable features that tended to convey success to the parents will be passed along disproportionately to their offspring. Should the rules change so that some other heritable feature comes to be the best fit to a modified environment, that bias in the transmission of genetic information will keep pace. Evolutionary change will be the natural result.

No one, least of all an ultra-Darwinian like George Williams, would dispute this simple characterization of natural selection. Natural selection in this familiar, traditional Darwinian guise remains that passive recorder of "what worked better than what." The collective genome of a population at any one moment is a status report reflecting what proved superior to what in the field of heritable variation of previous generations.

The ultra-Darwinian movement, beginning right after the Darwinian centennial of 1959, set out to change the view of natural selection as a passive process to a decidedly active one. True, as George Williams stressed (and all agree), selection has no "eyes" for the future. There is nothing in nature that parallels the breeder, who aims to fluff out the coats of a strain of sheep. How then to put natural selection in an active sense, not coincidentally rendering selection an active force in nature?

By 1966, the year of publication of George Williams's *Adaptation and Natural Selection*, population geneticists had long since taken to thinking of natural selection strictly in terms of relative reproductive success—relative, that is, to other members of the breeding population. Selection had become a matter of an organism outcompeting its rivals in leaving relatively more copies of its genes to the next generation. Indeed, the very word *fitness* had changed from its general, near-colloquial original sense in Darwinian usage. Fitness to Darwin meant something like overall vigor. Those individuals best suited to cope with life's exigencies were the more "fit." The more "fit" were more likely to be the ones to produce more offspring, thus leaving more copies of their genes to the next generation. But, following the lead of Ronald Fisher and other early population geneticists, the defini-

tion of "fitness" became elided—now simply meaning "reproductive success." *Relative fitness*, in this new sense, is a measure
of relative reproductive success among individuals (bearing different alleles) in a population. Natural selection became differential fitness—strictly a matter of what leaves more copies of its
genes to the next generation.

Population geneticists realize that many factors influence
fitness. Cardinal among them is relative success in the ecological
world, where variations in a particular structure (say tooth
shape) have definite implications for survival, and thus for
reproductive success. As good adaptationists, Darwinians never
did forget the link to the external environment implied by the
old definition of fitness in the original evolutionary dialogues.

Nevertheless, linking natural selection so intimately with
reproductive activities set the stage for the next ultra-Darwinian
gambit. Williams, reiterating that selection cannot "see" the
future, concluded—not unreasonably—that organisms cannot
possibly be said to reproduce for the purpose of perpetuating
the population or species of which they are members. There is no
way natural selection can "know" what lies in store for a species
as time wears on. That being said, Williams simply asked what
end reproduction actually serves. He concluded that, if reproduction cannot be for the good of the species, *then it must be for
the good of the individual.*

According to Williams, the "goal of an individual's reproduction is . . . to maximize the representation of its own germ
plasm, relative to that of others in the same population." These
are strong words. They are the very underpinnings of High Table
ultra-Darwinian evolutionary biology.

But Williams did not stop there. Because fitness—defined
as reproductive success—is in large measure a reflection of an
organism's *economic* success (for example, in obtaining food), an
organism's reproductive goal underlies its other, more immediate goals. That is, success in the hunt is likely to confer success in
mating. In a striking passage, Williams wrote of the separate
goals of a fox and a rabbit, each out to "contribute as heavily as
possible to the next generation." And, while neither is out to
frustrate the goal of the other, nonetheless "the achievement of

the fox's goal may require, at the tactical level, the death of the rabbit."

From an ultra-Darwinian perspective, then, an organism eats in order to further its reproductive goals. It might eat, first, to live; but it lives, from this perspective, in order to reproduce. All actions an organism takes—whether directly reproductive, or simply concerned with the tasks of gaining energy and avoiding predation—are carried out with the competitive purpose of leaving a greater share of genes to the next generation.

And thus has the concept of selection been utterly transformed. No longer a mere passive ledger recording which variations worked the best in the preceding generation, *selection is now seen as the direct outcome of active competition between individuals in a population in order to leave more copies of their genes to the next generation.* And the way is now paved for interpreting all manner of organismic activity as devices to forward this competitive race.

Once the notion that all the activities of organisms could be read as a form of competition for reproductive success was established, it was not long before the next step was taken in the form of Richard Dawkins's *The Selfish Gene* (1976). According to Dawkins, there are in life but two kinds of entities: *replicators* and *vehicles* (a terminological distinction from Dawkins's *The Extended Phenotype* (1982)). Genes are replicators, but they can't exist and operate on their own. Genes need a vehicle—an organism— to house and nourish them and to facilitate their replicative functions. It isn't organisms (the mere vehicles) that are competing for reproductive success, but the genes themselves. Organisms as vehicles are simply the unwitting dupes of their genic components.

The Selfish Gene tells us that the instructions for building a system—an organism, and by Dawkins's own extension, social systems and ecosystems—are in a very real sense more important than the system itself. The organism exists merely to aid in the spread of genes. The egg came before the chicken and is always to be construed as the more important of the two.

But I'm old-fashioned. In the post–Watson-Crick enthusiasm for the uniqueness of genetic systems in the universe, we tend to overlook the other great facet of biological systems: the mundane fact of material, corporeal existence of organisms as matter-energy transfer systems. Tradition would have it that these systems exist in their own right—and here I am an ardent traditionalist. Though it is unfashionable these days to think of organisms as machines, to me Williams's fox is a matter-energy transfer machine that needs to catch and devour Williams's rabbit merely to continue to exist. Williams thinks the fox eats the rabbit in order to pass on its genes. Dawkins would call Williams's fox a vehicle for its own genes' desires for replication. I, to the contrary, see reproduction as a physiological luxury rather than an imperative that is necessary for that fox to go on living. In my view, if and only if that fox's economic life is going well can it afford to reproduce.

Is it fair to say that life is "all about" reproduction and the handing down of genes from one generation to the next? If reproduction is not for the "good of the species" must we then conclude that it is necessarily for the good of the individual organism? To ultra-Darwinians, the answer to both questions is an unequivocal "yes." To them, this translation of the Darwinian notion of natural selection explains not only how evolution happens, but the very nature and structure of all manner of biological systems—from breeding systems within populations to entire species and ecosystems.

That, then, is what adaptations are to ultra-Darwinian eyes: devices to facilitate the spread of an organism's genetic information—a process which benefits the organism itself. We'll return to this posture time and again as the discussion unfolds. But we should bear in mind that George Williams was out to improve the scientific study of adaptation—which he called an "onerous concept"—when he formulated this active conceptualization of natural selection. Nor has Williams been the only biologist to suggest ways in which the study of adaptation can be rendered more rigorous, more "scientific." Among them are my colleagues in the naturalist camp, Elisabeth S. Vrba and Stephen Jay Gould. Their work, too, has meant to clarify the nature of adaptation. Some of it has managed to infuriate the ultra-Darwinians.

Dr. Pangloss

George Williams was not alone in his distaste for the rather careless approach to adaptation seen in traditional evolutionary biology. Students of adaptation had long since fallen into a rut. Their analytic framework was always the same: a study of how some structure works was followed by a flat-out assertion that the complex functioning of structures are carefully fashioned through natural selection. For instance, a woodpecker manages to blast into a tree with such rapidity and force without scrambling its brains because the bones and muscles of a woodpecker's head are built through natural selection precisely to avoid brain scrambling. The functional anatomical analysis would often be elegant, but the evolutionary homily at the end, presented as a "conclusion," was in reality nothing but a statement of the underlying assumptions brought to the research in the first place.

In a sense, organismic biologists—meaning nongeneticists focusing on the structure, function, classification, and ecology of animals and plants—were in a trap when it came to evolution. After all, the synthesis had decreed that all evolutionary mechanics were to be construed as dynamics of gene distributions within populations, thus strictly the province of genetics. The "Just So Story" evolved as a statement of principle—of rote application of the black box of natural selection in order to explain the origin of a particular function, structure, or item of behavior. The stories were probably on the money, more often than not. But there is no way of knowing for sure, and the entire exercise of adaptive storytelling began to strike many evolutionary biologists as inherently dissatisfying.

Thus, there was a great outcry against "just so stories" heard in many different quarters of evolutionary biology from the 1960s right on through to the present day. Rudyard Kipling's famous fables recounting how the elephant got his trunk, the rhino his wrinkly skin, and the leopard his spots had the advantage of great style and manifest whimsy. But no one would mistake his yarns for historical truth, the more so that they universally relied upon the long-discredited mode of the inheritance of acquired characteristics that had been expunged from biology since the days of August Weismann at the end of the nineteenth century. Even so, the Darwinian *adaptive* story of how giraffes got their long necks were often little better (and considerably less elegant) than Kipling's narratives. They were only stories about how natural selection may have acted to produce longer necks so that giraffes might take advantage of the leafy greens in tree canopies which were unavailable to antelopes and other African herbivores.

Plausibility and consistency with the basic tenets of population genetics provide the sole criteria for evaluating these historical reconstructions. In the typical just-so evolutionary tale, there was little of the "rigor" justly so revered by scientists. There was no way, in other words, to verify the truth of any one version—to "test the hypothesis." Small wonder so many evolutionary biologists yearned to see the study of adaptations sharpened and brought up to scientific snuff.

If most evolutionists could agree on the wisdom of a more rigorous approach to adaptation, one particular event both deepened and polarized this issue around the High Table. In 1979, Maynard Smith convened a meeting on adaptation and natural selection. As he himself acknowledges, the paper that had by far the greatest impact was contributed by Stephen Jay Gould and Harvard evolutionary geneticist Richard Lewontin.

Gould and Lewontin went far beyond the just-so critique that was already cropping up all over evolutionary biology. They charged that a fixed mind-set had pervaded evolutionary biology for the past 40 years—an "adaptationist programme . . . based on faith in the power of natural selection as an optimizing agent." Voltaire's fictional Dr. Pangloss tended to see a purpose for everything in "this best of possible worlds" ("Everything is made for the best purpose. Our noses were made to carry spectacles, so we have spectacles..."). So too, said Gould and Lewontin, with natural selection honing adaptations—in the minds, that is, of evolutionists caught up in what they called the "adaptationist programme."

In this view, selection constantly seeks to optimize adaptations, and departures from optimality come only when selective conflicts arise. For example, if the name of the evolutionary game, in ultra-Darwinian terms, is to maximize one's own genetic representation in the next generation, selection would naturally work to optimize the number of offspring produced in each generation. As a practical matter, however, there is a limit to how many offspring can be successfully reared to maturity. In the case of birds, for each different species, there is an upper limit on the number of hatchlings that can be successfully reared. Too many hatchlings in a nest can overtax the parental ability to feed them, and fewer young survive than had there actually been fewer eggs to start with, or so goes the argument. There's an implicit trade-off, a compromise reached between producing too many, or too few, eggs. And that compromise is in itself a form of optimality—honed by natural selection.

No one disputes this egg-optimality argument. But, Gould and Lewontin asserted, errors are bound to arise when absolutely all attributes of organisms are treated in this fashion. The danger is not so much in making up uncritical (and inherently uncriticizable) "just so stories," as it is a systematic and willful refusal to examine alternative reasons organisms have come to look and behave the way they do. Not every feature need be thought of as a masterpiece of natural design, they argued—and there is a lot more going on in evolution than just the process of natural selection maximizing adaptations.

Gould and Lewontin, citing Darwin's original pluralism (all of us, at one time or another, seek to bolster our argument by claiming a pure Darwinian pedigree), pointed to a number of factors simply left out of neo-Darwinian optimality arguments. Chief among them was the observation that, in the urge to explain the functional parts of organisms in terms of their underlying genes, somehow the organism itself, the living being, has been left out. Think of the intricacies of developing a complex animal, with its billions of cells, hundreds of cell types, and dozens of tissue types, not to mention a score or more of organs, all from a single fertilized egg. These intricacies must severely limit the capacities of selection to alter fundamental aspects of organismic design. There must be relatively few ways in which the developmental process can be altered and still yield a viable, adult organism. Such developmental and architectural "constraints," argued Gould and Lewontin, are as important to evolutionary history—governing the actual direction that evolution can take—as instances where selection does manage to alter developmental pathways in the evolutionary transformation of adult anatomy from one form to another. Ultra-Darwinians grasp the point, but see the Gould–Lewontin emphasis on developmental constraints as overblown.

Moreover, Gould and Lewontin charged that the "adaptationist programme" simply leaves out alternative explanations of evolutionary change. Sewall Wright had long since pointed out that some alternative forms of genes (alleles) are expected to become either established or lost in populations on a purely random basis. And some features seem to tag along perforce, by-products of the evolution of some other trait. The simplest and clearest examples come from the study of allometry (differential rates of growth of different body parts)—one of Gould's early topics of research. Titanotheres, which are extinct relatives of horses and rhinos, offer a classic case history. The earliest titanotheres were roughly the size of ponies. As time went on, however, larger species of titanotheres evolved, culminating in some massive giants, among the largest land mammals ever.

The earliest, relatively small titanotheres had small bony bumps on the head. Not surprisingly, larger titanothere species came to have larger horns. But the heads of the bigger species were not just scaled-up versions of their smaller, more primitive kin. Rather, the bigger the head, the relatively bigger the horns became. As time went by, the horns grew faster than the rest of a titanothere head. And, while it is possible that selection was acting to produce progressively larger titanothere horns, Gould and Lewontin's argument suggests that selection might only have been for increased body size. The disproportionate lengthening of the horns may well have been a side effect, a mere allometric corollary of the evolution of larger body size.

The adaptationist programme, with its penchant for focusing on body parts, has produced a rich field of adaptive scenarios for increase in titanothere horn size—all based on the assumption that relatively larger horns had to have been developed to serve some specific purpose. Adaptationists assume that larger titanothere horns are adaptations honed by natural selection. But they might be not be adaptations at all. They might simply be side effects of the real adaptation: increase in overall body size.

And then there was Williams's own point to consider: one cannot assume in general that natural selection shaped a structure or behavior to perform the particular function we now observe. No one argues that natural selection shaped the human brain, concomitantly fashioning its cognitive capabilities, for the specific purpose of making and appreciating music. Gould and Lewontin reiterated Williams's argument as part of their characterization of the pitfalls of the "adaptationist programme."

But it was not until Elisabeth S. Vrba grappled with the problem in the early 1980s that the force of this criticism really took hold. Vrba argued that structures fashioned by natural selection to serve one particular function may well later be co-opted to serve yet another, and that it would be wrong to call the second function an "adaptation." If so, it would be wrong to claim that natural selection had shaped the structure to perform that secondary, co-opted function.

Gould and Vrba, examining these problems in detail in a 1982 paper, offered a particularly graphic example of this phenomenon of the co-option of structures to perform functions unrelated to the original evolution of the structure. The African black heron uses its wings to cast a circular shadow on the shallow water in front of it, enticing fish to a shady place right under its expectant dagger-like beak. No one would claim that the heron's wings evolved to delude fish as part of its feeding behavior. Wings, instead, are obviously "for" flight—although here we heed the cladist credo and recognize that wings did not evolve *de novo* in the African black heron: they were inherited, in only slightly modified form from its ancestral heron species.

Indeed, there is even reason to think that back in the Jurassic, in the ancestral species of dinosaurs that gave rise to the bird lineage, wings may not have been developed originally for flight at all. Flight itself may have been a co-opted function of feathered wings, which may have originally been fashioned by natural selection for heat regulation, or even for trapping insects on land (not terribly unlike the secondary use to which modern African black herons put their wings).

In any case, Gould and Vrba pointed out that a structure that performs some specific, useful function should not automatically be termed an "adaptation." It is, rather, an *aptation*. If shaped by natural selection to perform its present function, it is properly called an "adaptation." If not, as in the case of the black heron's fishing behavior, or of humans making music, it is co-opted, an "exaptation."

Throughout his writings, including his monthly column in *Natural History* magazine, Steve Gould has stressed a few simple messages. One is that organisms are not so much paragons of design as compromises of design. Against the ability of natural selection to hone adaptations to perfection and to fashion organisms maximally suited in every way to the ecological worlds in which they live, stand several formal obstacles. For one thing, genetic variability (on which natural selection works) may simply not be available. Then there is the dead hand of history: further evolution depends very much on what it has already produced, limiting design opportunities for the future. This is as important a constraint on evolution as developmental conservatism.

Then again, there are the limitations imposed by architectural design itself. The spandrels of San Marco (that appear in the title of the Gould–Lewontin paper on the "adaptationist programme") are "tapering triangular spaces formed by the intersection of two rounded arches at right angles"; these are "necessary architectural byproducts of mounting a dome on rounded arches." Because of their prominence in church decorative arts, it might be tempting to judge the spandrels differently—to conclude that they are there to provide critically important space in which to illustrate major Christian themes. But the truth is that they are there simply because they have to be, and early church decorators simply took advantage of their presence.

High Table ultra-Darwinians, while conceding the need to clean up the adaptationist storytelling act, are generally unmoved by the Gould–Lewontin call for pluralism. They like to paint Gould as the Anti-Darwin, a man who has renounced adaptation as the central facet of evolutionary science. It's a point of rhetorical convenience for ultra-Darwinians to take this stance, simply because selection-mediated adaptive evolution forms virtually their entire subject matter. To them, that is absolutely all there is to evolution. It helps their cause to paint a chief critic as one who has utterly abandoned the concept of adaptation through natural selection—an absurd position no serious evolutionary biologist, starting with Steve Gould, would ever adopt.

Maynard Smith thinks it would have helped if Gould had been, like him, a bird-watcher rather than an aficionado of dinosaurs. Birds do things. So did dinosaurs, of course. But with their scattered bones now frozen in the rocks, it takes an extra effort to think of them as living beings, actively engaged in finding food, fending off predators, and mating: the very things all those evolutionary adaptations are designed to do.

This complaint, however, won't wash. True, Gould openly disdains bird-watching. *Chacun a son goût.* I happen to love it, and have even spent a few pleasant hours with Maynard Smith as he kindly tried to show me some European songbirds. The experience hasn't made me an ultra-Darwinian. Beyond the sheer pleasure of it, birding does (as Maynard Smith says) reveal endless details of bird behavior that make you wonder why they are doing what they are doing. But, for me, watching birds does a lot more. It makes me think about what species really are, what holds them together, and what roles species themselves might play in the natural world. Bird-watching makes me wonder why so many of the birds I see so closely fit their picture in the field guide—in apparent opposition to the maxim that natural populations are rife with variation. Birding makes me think of the roles of different species in local ecosystems. It makes me wonder about the context of adaptive change. How is adaptive change related to the appearance of new species? What factors act to keep species stable for long periods? What happens when adaptive change does occur—often, if not always, quite rapidly?

Ultra-Darwinians, then, tend to agree that much of the adaptive storytelling in evolutionary biology needs to be sharpened. But they won't budge on the larger point: that there is more to the history of apparent organic "design" than meets the eye. They remain confident that natural selection has shaped most anatomies and behaviors to perform specific functions. And they remain averse to reverting to pluralism in evolutionary explanation—vastly preferring the narrower focus of adaptation through natural selection that had emerged triumphant and alone by 1959.

For my own part, I am content to let this aspect of the debate simmer. Whatever the relative roles and magnitude of impact constraints and the laws of relative growth might have had on the evolutionary history of organic form and function, none of us doubts that the main signal in evolutionary history has indeed been contributed by adaptation through natural selection. I prefer to confront ultra-Darwinians, not so much on the issue of the actual importance of selection-mediated adaptive change, but on the details of *how and under what circumstances* selection acts to cause evolutionary change.

Before turning to these issues, however, we need to confront one lingering point: how, indeed, can we inject some rigor into the analysis of adaptation? For there has indeed been progress. It comes from a somewhat surprising quarter—a cadre of systematists, many of whom deny any overt interest in the evolutionary process whatsoever.

Cladistics: A Search for Rigor in Evolutionary History

Creationists are fond of claiming that no one can say anything definitive about evolutionary history, simply because there were no humans back in the Cambrian Period to record the early stages of life's evolution. A not altogether dissimilar claim, oddly enough, permeated the ranks of evolutionary biology until fairly recently. As a student in the 1960s, I was regularly informed that the only way to gain a glimmer of evolutionary history is to have at hand a dense, rich fossil record.

As a budding paleontologist, this was not necessarily bad news. If indeed the sole function of paleontology is to record the very fact of evolution, it was altogether fitting that we paleontologists should be the sole keepers of the evolutionary history of life. Nice to have something to do; and poor as the record is for entirely soft-bodied creatures, and incomplete as it is even for organisms equipped with hard skeletons, few biologists questioned the conventional wisdom that only the fossil record could reveal the outlines of biological history.

But conventional wisdom presents an inviting target. Forgotten in all the post-Darwinian rhetoric was a clear signal from Darwin's own *On the Origin of Species*. Darwin had but one illustration in that seminal work, and he referred to it several times to make various different points. I reproduce that diagram on the next page, and reiterate the most important use that Darwin found for it: the most compelling evidence that life has actually evolved.

Imagine, Darwin said, a lineage existing through geological time. Imagine that, as time goes by, the characteristics of the organisms in that lineage should become modified through some natural process. Their descendants, naturally enough, will inherit those modifications. Now, imagine but one thing further: that as time goes by, lineages diverge, branching off from one another. Descendant branches will likewise inherit the modifications of their common ancestors, but once split apart as separate lineages, any further evolutionary modifications will only be inherited by descendants within each of the divergent lineages.

Now, how do these simple suppositions establish the very fact of evolution? Simple. Looking around at the more than ten million species currently on earth, we see a nested pattern of resemblance interlinking them all. And that is exactly what we would expect to see given Darwin's "descent with modification." If all life has descended from a single common ancestor,

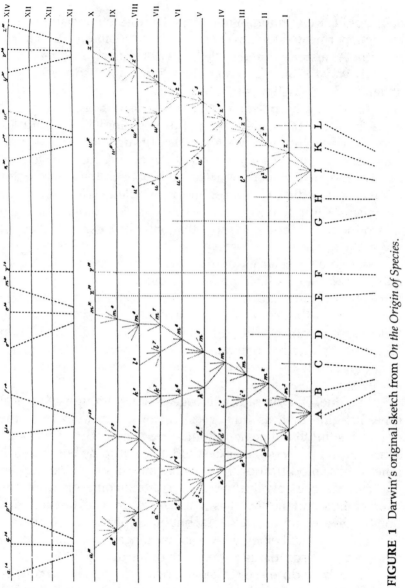

FIGURE 1 Darwin's original sketch from *On the Origin of Species*.

there should be some features that have been inherited by absolutely all living things. And that's what we see. Fundamental biochemical pathways and, perhaps more graphically, the macromolecule RNA is common to all forms of life (DNA is nearly as ubiquitous, but is not found in some bacteria).

Take any branch of life and you will see the same thing. We share hair, mammary glands, placentas, and three middle ear bones with other mammals; in our more remote past, we shared with birds and reptiles a special system of liquid-holding membranes in our eggs—the amniote egg. Reptiles, birds, and mammals share a four-leggedness with amphibians, and all these groups share a backbone with the various groups of fishes, and so on and on, back into the dim recesses of time, all the way back to all of life sharing its RNA.

This detailed pattern of resemblance linking absolutely every living thing is the very best evidence that life has indeed had a long evolutionary history. We see in this pattern the outcome of a simple, natural process. Evolution is a thoroughly scientific proposition: if evolution is "true," we would predict that such a pattern must be there. It is.

The nested pattern of resemblance ought to be at the very top of the list of the "proofs" Darwin enumerated in the *Origin*— a list required in all decent high school biology curricula. Oddly enough, though, it is seldom stressed, even by professional biologists. In retrospect, it is ridiculous to think that only paleontology can tell us anything about evolutionary history. In terms of pure genealogy, the living world carries the marks of its history around with it.

Biologists, in other words, can reconstruct the genealogies of living species in a "who is more closely related to whom" sort of way—even when many of the intermediate relatives have long ago become extinct. We share an amniotic egg with birds and reptiles, so we know that we must have shared a common ancestor with those rather different-looking lineages sometime in our evolutionary past. We have some additional hints that this must be so. The duck-billed platypus has hair, but lacks a placenta, and so is more reptile-like than we are (its "duck-bill," however, is not really the same structure as a bird's beak!). And, sure enough, the fossil record shows an abundance of "mammal-like" reptiles 250 million years ago. These primitive mammals, members of our own lineage, still retained many of the features of our common ancestors with true reptiles and birds.

Systematics is the branch of evolutionary biology charged with discovering, naming, and classifying the world's species. Pre-Darwinian systematists saw the nested pattern of resemblance in the living world, and sought to capture this "natural order" in their classifications. Rather like the periodic table of the elements, classifications sought to place species where they "belonged." After Darwin, everyone saw that "where they belonged" meant "who was most closely related to whom."

A great surge of enthusiasm for tracing pedigrees of plant and animal groups gripped late nineteenth-century biology. But systematics was part of the eclipse suffered by evolutionary biology in general with the rise of genetics at the turn of the century. With the triumph of neo-Darwinisn and the rise of evolutionary synthesis in the late 1930s, though, systematics made a comeback. England's Julian Huxley called it the "new systematics." Its theoretical underpinnings were the evolutionary postulates of the modern synthesis itself. As the synthesis "hardened" into a nearly pure form of adaptation through selection, systematics followed suit.

But how to reflect the history of adaptation in a classification? Should a systematist concentrate just on shared adaptations? Or should evolutionary uniqueness also enter in? Bernhard Rensch, for example, proposed that we humans, species *Homo sapiens*, be placed in our own Kingdom ("Psychozoa") in recognition of our highly specialized and evolutionarily cognitive capabilities. This would remove us from the Kingdom Animalia! Genealogically, of course, we are kissing cousins to the great apes, and biologists these days classify us in the same family with chimps, gorillas, and orangutans.

Systematists imbued with the adaptationism of the synthesis had veered far away from pure genealogy. Simpson was inspired to remark that classification is really as much an art as it is a science—words that were to haunt him in his later days. It was the German entomologist Willi Hennig who started the movement back to basics, redirecting the systematist's gaze to the actual signal of relatedness latent in the anatomies, physiologies, and behaviors of the world's living species. He called his system "phylogenetic systematics." It was nothing less than a set of rules on how to analyze the pattern of resemblance that we have just seen actually does link all living species. That pattern is there, and armed with a well-conceived analytic protocol, there is no need to think of systematics as an art form.

Ernst Mayr and George Simpson bitterly opposed this encroachment—though each eventually conceded that, in terms of pure genealogical reconstruction, Hennig's system made great sense. Mayr coined the term "cladistics" as an unflattering reference to the emphasis solely on genealogical lineages ("clades") with no attention to other aspects of adaptation.

Cladistics caught on like a house afire. Especially through the efforts of Gareth Nelson, a young systematist at the American Museum of Natural History who argued vociferously the merits of Hennig's system, a bunch of us jumped on the bandwagon. Precisely because it works, and because it has provided systematists for the first time with a rational set of working principles, cladistics is now firmly established the world over as the way to perform systematics research.

Part of the beauty of cladistics is that it does not rely on a set of precepts about how the evolutionary process works. All you need do is see that nested pattern of resemblance connecting up the living world, and then reconstruct the genealogical patterns by looking for those items of special resemblance (like hair in mammals) that link species together. There was no excess conceptual baggage, especially the strong rhetoric about adaptation that had permeated systematics at least since the 1930s.

Ultra-Darwinians were slow to see the virtues of cladistics. They openly ridiculed cladist proclamations of independence from evolutionary theory—and Maynard Smith still sometimes refers to cladistics as a "religion." Yet, ironically enough, for all the antievolutionary rhetoric associated with some cladists, it is cladistics that has actually placed the study of adaptation itself on a firmer scientific footing! Even Maynard Smith thinks so.

And here's why. Evolutionists have always known that convergence—the development of similar structures independently in different lineages—is the very best evidence for evolutionary adaptation. Organisms may resemble one another because they have inherited their features from some common ancestor. However, such evolutionary resemblance is not good evidence of adaptation in any one species. But if the resemblance could be shown to be superficial, not inherited from a common ancestor, the case for evolutionary adaptation underlying the appearance of a structure or function is much more credible. For example, the fusiform shape of sharks, porpoises, extinct ichthyosaurs, and tuna, in functional terms, renders a streamlined animal well suited for a rapacious life of fast swimming carnivory at sea. Biologists long before the advent of cladistics saw that sharks are primitive cartilaginous "fish," not terribly closely related to bony fishes (like tunas), and that ichthyosaurs are reptiles, and porpoises are mammals. It would be a mistake to classify these disparate organisms together because of their superficial resemblance. This is the very essence of cladistics: to erase similar, harder-to-detect mistakes.

Take, for example, arthropods—by far the largest phylum of complex animals. Biologists steeped in the modern synthesis in the late 1950s and 1960s became entranced with the idea that the arthropod skeleton—a hard, outer casing that must be shed for growth to occur—is such a handy, functional arrangement that it must have arisen not once, but several different times. Arthropods, in this view, would not be a natural evolutionary group.

Convergence is not unknown in arthropods. Spiders and their kin, insects, and terrestrial crustaceans (isopods—"sow bugs") are land-dwelling, air-breathing animals. Spiders and isopods each have their closest relatives living beneath the waves, using gill structures to extract oxygen from seawater. That all three land-dwelling groups have lunglike structures communicating to the outside through tiny holes in the skeleton is a bona fide case of convergence (and an excellent argument of adaptation for air breathing enabling life on land).

But the tools of cladistics even now are grappling with the larger issues of arthropod evolutionary history. Are all arthropods descended from a single common ancestor, or are they a case of convergent evolution on a truly grand scale? Only a cladist can tell, and the evidence these days is mounting that arthropods are truly monophyletic (that is, represent one single evolutionary lineage). Suppositions that arthropods arose more than once are just that, speculations based to no small degree on the assumption that adaptation through natural selection—rather than simple genealogy—accounts for a grander share of resemblance among organisms.

Cladistics, then, serves as the arbiter in conflicting claims: does adaptation or simple inheritance underlie close resemblance between species in nature? Cladistics does much to take the guessing, the "just so story" scenario out of studies of adaptation. And evolutionary biologists are getting the point loud and clear. A number of studies have appeared in recent years, studies that use cladistics to take the guesswork out of what is an adaptation and what is not. Precision in working out the details of the history of life has injected a much needed form of precision to the actual study of adaptation. And that's an irony, because adaptationism had to be excised from systematics before it could become of any use to evolutionists concerned with adaptation!

Coda: The Limits of Adaptationism

When lecturing to new audiences, I like to present myself as a "knee-jerk" neo-Darwinian, at least when it comes to the matter of adaptation and natural selection. It's true enough, and comes as something of a surprise to some who suppose that I will promulgate some wild new theory to supplant traditional canon. People tend to equate punctuated equilibria with some alternate notion of how evolutionary change—adaptive evolutionary change—occurs.

But no rational evolutionary biologist feels that most change is not adaptive, or that adaptive change is not caused by natural selection. As we begin to see in the next chapter, our fossil data imply that evolutionary change is rather more difficult and rare than generally suspected. And when it does happen, adaptive change comes rather quickly, at least when compared to vastly longer periods when species don't seem to change all that much.

But what happens during those much briefer spurts? Adaptive change through natural selection, that's what—in the context, to be sure, of true speciation. And though our data frequently are too poor to demonstrate gradual change through selection, we do in fact have some documented examples of smooth transitions that are very much in agreement with natural selection.

That being said, I confess that as a student I disliked the prospect of spending my life rediscovering the wheel—of looking at adaptive change (and stability) and proclaiming such to be further evidence of adaptive evolutionary change through the aegis of natural selection. I saw little prospect of making any further contribution to understanding evolution. Focusing on adaptation per se, I would be consigned to a mop-up role, merely documenting the fact of evolutionary history, applying a rote interpretation to my fossils' reconstructed evolutionary histories. Or so I worried, not unrealistically. No real fun in that, I thought—especially since I am not a particularly gifted anatomist.

So I frankly admit I started to look for something more, some other way to approach the problem. I stumbled on evolutionary patterns in the grand scale of geologic time, patterns involving species as well as the anatomies of individual organisms. It has, I believe, opened up new vistas of the true arena, the real context, in which the evolutionary game is played.

3

The Great Stasis Debate

Science is a search for ever-sharper images of physical reality. We need to know what kinds of things populate the material universe: what they are and what kinds of things they do. Much of the debate around the High Table is over just that list of "things"—biological systems. Ultra-Darwinians restrict their list pretty much to genes, organisms, and populations—acknowledging that species, social systems, and ecosystems exist, but not as direct players in the evolutionary arena. In contrast, I see such large-scale systems as absolutely crucial to understanding how the evolutionary process actually works.

Over and above the debate on the nature of biological systems, though, are more objectively determinable, empirical claims about the nature of things. Empiricism is the traditional touchstone of science, and there is no doubt that the more concrete phenomena are, the more readily they can be measured and definitively characterized, the more sure we are of what we are talking about, and the more likely we are to reach some measure of agreement. Though it must be said that scientists are as prone to argue over the "facts" of a matter as over their interpretation!

In any case, we naturalists present as our opening gambit and our strongest suit what we take to be the pertinent "facts" of the matter: patterns of evolutionary history. By "patterns" I simply mean that the evolutionary histories of lineage after lineage bear a haunting, specifiable familiarity about them. Think of the fossil record as the result of literally millions of evolutionary "experiments"—experiments run in evolutionary real time, but without the controls that laboratory scientists bring to their experiments in order to facilitate interpretation of the results. But experiments they are—and it stands to reason that we, following Simpson's suggestion a half-century ago, use these results as a yardstick to compare rival notions of how the evolutionary process actually works.

It is a two-step process: first, establish the very existence of a recurrent pattern in evolutionary history. Then, using what we know about the evolutionary biology of living organisms, formulate a combination of factors that is most likely to yield those very kinds of evolutionary patterns we see in the fossil record. That was Simpson's original gambit. The only alternative is to sit in an armchair and imagine what evolution might look like over vast segments of time, based on the short-term patterns we can elicit in laboratory experiments. Computer simulations, which can address large numbers of generations over prodigious amounts of time, are nonetheless only as good as the assumptions built into them. Extrapolation from the short term to true evolutionary time is not the only game in town. And it is time to see how poorly traditional Darwinian extrapolation fares when held up against the light of the fossil record.

Evolutionary Pictures in the Mind's Eye

Imagination lies at the heart of scientific endeavor—just as in any other creative human enterprise. So it is no sin to imagine what evolution might look like over truly long periods of time based on observed generation-to-generation change. What has been missing is the second half of the equation: comparing those pictures of change with actual longer-term observations we have made—observations largely if not exclusively from the fossil record. As we shall see, when we do make such a comparison, the traditional picture is stunningly out of whack. To see just how off base it has been, we first need to take a closer look at what the standard expectation of evolutionary change has been, and continues to be.

Ultra-Darwinians are ultraconservatives when it comes to imagining what evolutionary change actually looks like in geological time. In fact, theirs is essentially the same mental image of evolutionary pattern as that first championed by Darwin in the mid–nineteenth century.

Darwin, recall, was out simply to establish the basic fact of evolution. He succeeded where others before him had failed in no small measure because he could specify an engine of evolutionary change—natural selection. Selection, he argued, would act to perfect the fit between organism and environment while environmental conditions remained stable. Should those conditions change, and if there were enough heritable variation in the population, natural selection would then adjust a population to this new altered set of environmental circumstances.

By the mid–nineteenth century, it had already become clear that major environmental change was the rule, not the exception, in earth history. Ironically, a staunch opponent of the very idea of evolution, the Swiss biologist Louis Agassiz, was the one who produced the most striking piece of evidence that climates can and do change radically over the course of time. Agassiz established that Europe had undergone a series of prodigious glaciations in the comparatively recent geological past. Vast, thick sheets of ice had pushed down from the polar regions, only to retreat, then come once again—in an oscillation that has happened (we now know) four times, so far.

Darwin depicted a world in flux. According to his picture, environmental change is inevitable; therefore, the adaptive response of species to this change is itself inevitable. The result is—Darwin imagined—that as a consequence of their existence in a changeable world, species will inexorably and inevitably keep changing, always evolving in response to ongoing environmental change.

The American geneticist Sewall Wright pictorialized this basic Darwinian image of gradual, progressive change in such graphic terms that it remains the dominant metaphor of evolutionary change to this very day. No textbook—certainly including those written by ultra-Darwinians—lacks this essential depiction of gradual evolutionary change.

Actually, Wright had something rather different in mind when he came up with the idea of the "adaptive landscape" in the early 1930s. Mindful that certain gene combinations are bound to be, as he put it, "more harmonious" than others, Wright chose to depict those "more harmonious" combinations—the ones that produced the more robust, classically "fit" individuals—on high peaks of a topographic grid system. Less harmonious gene combinations constitute the lowlands of the adaptive landscape.

As Wright saw it, evolution would try to maximize the number of individuals within a species with the more harmonious gene combinations. Crucial to the process is the way species are organized in nature—broken up into a number of semi-isolated populations. Wright's characterization of the internal structure of species remains the view that all ecologists and evolutionary biologists basically accept. Each colony (later called a "deme") would be expected to have a semi-independent evolutionary history. Within each deme, Wright theorized that genetic drift as well as natural selection would act to maximize individuals with the most harmonious gene combinations. Because each deme has a somewhat different genetic complement from its neighbors, the adaptive peaks—meaning the most harmonious gene combinations—would be expected to differ from one deme to the next.

Wright himself extended his metaphor, at times departing from his original view that it is the best gene combinations, housed in individuals, that occupy the adaptive peaks. In some passages, Wright said that entire demes, and even species, occupy a single "adaptive peak"—referring to the optimality of adaptation of an entire population (deme), rather than of different individuals within each population. It was certainly this broader sense that Dobzhansky adopted in his seminal *Genetics and the Origin of Species* (1937), when, quite apart from Wright's original intent, "adaptive peaks" became something nearly akin to ecological niches.

This is the basic concept of "adaptive peaks" still much in evidence today—including in the discussions around the High Table. Geneticist Richard Lewontin provided a classic example of this imagery in his paper on adaptation in a special 1978 issue of *Scientific American* devoted exclusively to evolution. (This issue opens with an essay by Ernst Mayr in which he proclaims that evolutionary theory is effectively complete save for the dotting of a few i's and the crossing of a few t's.)

In a classic diagram reproduced on the next page, Lewontin follows standard practice and depicts adaptive peaks in profile—just like a statistician's "normal curve." The basic idea is that those organisms whose adaptive configuration is optimal for their generation are near the peak of the curve, while those with less optimal adaptations lie somewhere down the sloping flanks of the curve.

As time goes by, and environments change, the optimal configuration shifts. It is imagined to do so gradually. Lewontin's figure shows a species tracking environmental change, forever trying to keep pace with an ever-changing environment. At one point, the environment seems to bifurcate; the lineage divides, one species continuing to track the original direction of environmental change, and the other tracking a new opportunity—which also begins to change through time. Though Lewontin was soon to write the well-known "Spandrels of San Marco" paper with Steve Gould, in which they criticize specific aspects of the "adaptationist programme," Lewontin displays no special discomfort with the traditional depiction of gradual adaptive transformation.

Paleontologists have also been enamored with Wright's visual imagery, and routinely have depicted steady, gradual evolution in their textbooks. When Gould and I wrote our paper (in 1972) coining the term "punctuated equilibria," and contrasting

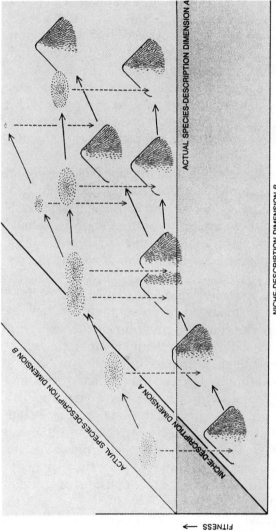

NICHE-DESCRIPTION DIMENSION B

FIGURE 2 Lewontin's diagram illustrating adaptation via the imagery of adaptive peaks. His original caption reads: "Species track environment through niche space, according to one view of adaptation. The niche, visualized as an 'adaptive peak,' keeps changing (moving to the right); a slowly changing species population (*colored dots*) just manages to keep up with the niche, always a bit short of the peak. As the environment changes, the single peak becomes two distinct peaks, and two populations diverge to form distinct species. One species cannot keep up with its rapidly changing environment, becomes less fit (lags farther behind changing peak) and extinct. Here niche space and actual-species space have only two dimensions; both of them are actually multidimensional." From Lewontin, R. C., *Adaptation*, p. 216, Copyright© 1978 by *Scientific American*, Inc. All rights reserved.

it with the standard picture (which we called "phyletic gradualism"), we used the Wrightian landscape imagery so common in paleontological texts as a major point of departure.

We chose the term "phyletic gradualism" advisedly. *Phyletic* (from *phylum*—a single branch or lineage) simply means that evolution—substantial amounts of evolution—is routinely expected to transform entire species as geological time rolls by. Lineages split, diverging in a process generally known as "speciation." In the standard depiction of such species bifurcations, each new branch marches off in its own gradual fashion. Speciation, in this traditional view, is just a special case of the general, gradual process of evolutionary transformation of entire species lineages through time.

The word "gradualism" is a bit more ambiguous. Gradual, of course, means "by imperceptible degree." But particularly in the early history of geological thought, it has also meant "slow" and even "steady." Sir Charles Lyell was a senior contemporary of Darwin, a geologist whom Darwin very much admired. Lyell's image of gradualism fit hand in hand with his famous principle of "uniformitarianism"—which itself (as Steve Gould pointed out years ago) means two distinct things. In one sense, uniformitarianism means that geological processes are just that: uniform from era to era. Thus we can interpret past events in light of present-day processes. But uniformitarianism has also meant uniformity of rate—a rather slow, steady rate, in contrast to the catastrophes that served as the focal point of rival geological theories in the first half of the nineteenth century.

Thus our term "phyletic gradualism" in general means "slow, steady change by degrees." In particular, it refers to the. slow, steady transformation of an entire species. We presented evidence that, contrary to the long-held picture of gradual evolutionary change through time, most species hardly change much at all once they appear in the fossil record—the phenomenon we called "stasis." We pointed out that paleontologists clung to the myth of gradual adaptive transformation even in the face of plain evidence to the contrary—paleontology's "trade secret," as Gould later called it.

On the other hand, it was not just because paleontologists hadn't yet forced geneticists to confront the phenomenon of stasis that neo-Darwinians—right on down through today's ultra-Darwinians—have tenaciously clung to the original Darwinian vision of gradualism. As we shall see, stasis is a simple prediction from Sewall Wright's much discussed work on species structure and evolution. Stasis is also implicit to some degree from empirical knowledge of what happens to species in the modern world as environments change. The ultra-Darwinian embrace of phyletic gradualism reveals some very serious flaws in their grasp of the basic organization of biological nature.

The Great Stasis Debate

If imagination is central to the scientific enterprise, so too, is common sense. Darwin had succeeded so well in overthrowing Whewell's dictum that "species have a real existence in nature, and a connection between them does not exist" that he managed to replace a worldview of static discontinuity with an expectation of flux. Species grade imperceptibly from one to another as the geological ages roll and environments inevitably change.

To the question, "What happens to species when environments change?", the standard post-Darwinian answer became, "They evolve." Species become transformed to meet the new conditions—provided, of course, they are well stocked with the necessary genetic variation on which natural selection may act to effect suitable evolutionary change. Failing that, the fate is extinction.

Here we have imagination colliding with common sense—and, worse, with empirical reality. Given the benefit of some 130 years of post-Darwinian scrutiny of the natural world, it has become abundantly clear that by far the most common response of species to environmental change is that they move—they change their locus of existence. In the face of environmental change, organisms within each and every species seek familiar living conditions—habitats that are "recognizable" to them based on the adaptations already in place.

This is "habitat tracking," the constant search for suitable habitat going on continually, generation after generation, within every species on the face of the earth. One hundred and twenty thousand years ago, lions and hippos roamed in what is now London's Trafalgar Square—graphic testament to much warmer climes in the period that separated the third and fourth massive continental glaciation events of the Pleistocene Ice Age.

On a much smaller scale, as the earth's climate has warmed steadily over the past century, biologists have charted subtle but persistent range changes as species adjust their locales accordingly. The Virginia opossum has been moving steadily northwards in the United States, as has the traditionally even more warmth-loving armadillo. Birds and insects have also been on the move, as have aquatic species. Plants, too: the staunch fixity of a tree is no barrier to migration, as seeds are dispersed and germinate wherever suitable habitat is encountered.

Granted, we enjoy the benefit of over a century's hindsight in documenting habitat tracking as the expected, usual response to environmental change. Even so, common sense and an ever-growing pile of ecological and paleontological data should long ago have replaced the Darwinian expectation of gradual evolutionary transformation—replacing it with the notion of habitat tracking.

Evolutionary biologists have always claimed that their canon fully incorporates the ecological perspective. More often than not, however, the incorporation amounts to a misreading of fundamental ecological principles. For example, species and larger-scale genealogical entities (higher taxa—such as Mammalia) are routinely said to occupy niches or otherwise play concerted ecological roles in nature. But species and higher taxa are purely genealogical entities. They are, in a very real sense, packages of genetic information. They do not live ensconced inside ecosystems, and thus play no concerted economic role in nature—a theme we'll encounter in greater detail later.

More recently, ultra-Darwinians have actually moved to subsume ecology within their evolutionary rubric—as when

Richard Dawkins announced in his *The Selfish Gene* that ultra-Darwinian principles, meaning competition for reproductive success—are what really underlie ecosystem structure and function. But in so saying, he simply ignores the dynamics of nonreproductive energy flow—the very hallmark of the ecological world that ecologists think are the actual organizing ingredients of ecosystems.

Naturalists, on the other hand, say that the most likely response of a species to environmental change is habitat tracking. The second most likely response to environmental change is extinction, which generally follows when suitable habitat cannot be found. The least likely outcome is wholesale, linear transformation of an entire species to meet the new environmental exigencies. Traditional evolutionists, including latter day ultra-Darwinians seated at the High Table, have it the other way around: environmental change begets evolutionary transformation through natural selection; failing that, we expect extinction. Habitat tracking doesn't even enter into the ultra-Darwinian picture—except to be airily dismissed as "fable," as George Williams has done in his recent *Natural Selection: Domains, Levels and Challenges* (1992).

As we shall see, adaptive change is very much tied in with environmental change, but in ways undreamt of in the ultra-Darwinian camp. The relationship between environmental change on the one hand and the collective processes of species extinction, adaptive change, and speciation on the other becomes clear only when we look at actual patterns of evolutionary history. We must do a far better job than traditional Darwinians have done in incorporating a realistic ecological perspective into our evolutionary theorizing.

Actual details of evolutionary pattern should have told everyone all along (Darwin himself included) that wholesale transformation of entire species is just not the way that adaptive evolutionary change manifests itself in nature. Some of these patterns are manifest in the living world. Sewall Wright's vision of the internal structure of species should have tipped everyone off that the wholesale transformation of entire species is not inherently very likely. But I do acknowledge that the patterns that most persuasively show that species do not routinely evolve themselves gradually beyond recognition come from the fossil record. And here I cannot fault the evolutionary biologists who—right on down through the modern ultra-Darwinians—are not privy to the details of paleontology. It would quite naturally be up to paleontologists to make this case. Why did we wait so long to do so?

In a sense, we didn't wait at all. In his book, *Darwin and His Critics*, philosopher David Hull has reprinted all of the reviews of any consequence of Darwin's first edition of *On the Origin of Species*. Among them are five written by paleontologists. Each of them expresses concern that Darwin does not acknowledge the by then well-known generalization that *once a species appears in the fossil record, it tends to persist with little appreciable change throughout the remainder of its existence.*

Thomas Henry Huxley alluded to this phenomenon of great species stability when he chastised Darwin for stressing the adage "nature does not make jumps" too strenuously. Because species do not, as a rule, grade imperceptibly one into another as we collect fossils up a cliff face, it follows that evolutionary change between species in the fossil record often appears to be rather abrupt. And Darwin himself, in his sixth edition (the usual version available in paperback form), acknowledged the tendency for species to remain stable throughout their duration in the fossil record. This was part of his strenuous and largely effective campaign to mute critics of his earlier editions by answering all their objections. (Darwin even allowed the inheritance of acquired characteristics to creep into his argument in the sixth edition. He was so anxious to preserve his fundamental thesis that life has evolved that he even relented on the issue of the total supremacy of natural selection as the shaper of organic form.)

Why, if Darwin himself acknowledged stasis in his sixth edition, was stasis not incorporated into the domain of standard evolutionary phenomena? It is clear that Darwin himself, even as he acknowledged stasis, did not stress it as the dominant signal of evolutionary history. He vastly preferred his original picture of gradual, even progressive, change. Certainly that was the picture that stuck most clearly in the minds of his contemporaries. I have already commented on the tendency of mainline evolutionary theory to stress continuity in the face of occasional assaults of nonorthodox discontinuity. Nonorthodox, that is, until Dobzhansky and Mayr were able to establish a discontinuity beachhead of sorts with their arguments on the nature of species and speciation in the late 1930s and early 1940s.

In any case, continuity, expressed as gradual transformation of entire species in geological time, was Darwinian orthodoxy from the very beginning. And paleontologists, I am sorry to say, quickly succumbed, or so it seems to me. The five paleontological reviewers of Darwin's original *Origin* appear not to have made any real impression on evolutionary discourse through their mention of the ubiquity of stasis.

In any case, the issue of stasis seems to have been quietly and rather quickly dropped. Stasis is far from a prominent theme in the paleontological literature of the last few decades of the nineteenth century. Thus Steve Gould's "paleontological trade secret": each generation of paleontologists becomes aware of stasis, only to shy away from it. Paleontologists, especially those devoted to marine invertebrate fossils, which typically leave far richer records than terrestrial vertebrates, have traditionally focused on the geological and environmental aspects of their fossils. They spend their time in the vital—though nonevolutionary—pursuit of using fossils to tell geological time, correlating fossiliferous rocks around the world according to their relative geological age. They also interpret the environments in which the ancient sediments were deposited—using fossils as ecological markers. Invertebrate paleontologists have for the most part eschewed any attempt to grapple with ideas of how life evolves; and, until recently, they have avoided systematic analysis of the patterns of evolutionary change, stasis, and extinction typically encountered in the fossil record.

For the most part, it has been paleontological reluctance to cross swords with Darwinian tradition that accounts for the failure to inject the empirical reality of stasis into the evolutionary picture. Indeed, ultra-Darwinian ears perked up after I resurrected stasis as a core ingredient of evolutionary pattern—the empirical essence of punctuated equilibria. The general idea (as expressed by Maynard Smith and also George Williams) is that stasis is an important phenomenon which, if a truly general evolutionary pattern, was not predicted by the formulae of population genetics. It was stasis that really got paleontologists back to the High Table.

If nothing else, convincing ultra-Darwinians of an important, unanticipated (by them), and little-understood evolutionary pattern has itself been a source of great satisfaction. True to High Table form though, we are far from agreeing on what actually causes stasis.

Indeed, all has not been entirely smooth even in the simple act of resurrecting stasis as an important evolutionary phenomenon. The contrast that I had in mind in the early 1970s—the very concept we still embrace today—is that species, far from regularly and systematically "evolving themselves out of existence," stay relatively unchanged throughout the duration of their existence. We knew, and openly acknowledged in our early work, that species exhibit variation. Indeed, there is variation within each local population. Also, just as in most modern species, there is variation between local populations spread throughout the total geographic range of extinct species.

The modern version of the old Darwinian picture of gradual phyletic change sees the field of variation present at any one time within a species translated and transformed through time. The typical scenario sees the average value of a particular variable trait shifting through time—as in Lewontin's pictorial account of adaptive peaks. Indeed, anatomical traits do shift around a bit as time goes by. But rarely do we see progressive transformation in any one direction lasting very long. What we see instead is oscillation. Variable traits usually seem to dance around an average value.

Typical are the results recently obtained by my colleague and former student Bruce Lieberman, who analyzed the evolutionary histories of two species of brachiopods (archaic shellfish) over a six- to eight-million-year interval that began roughly 380 million years ago. Both species (*Mediospirifer audaculus* and *Athyris spiriferoides*) changed a bit as time went by. But, after six million years of evolving, both species ended up looking very much like they did when they first showed up in the fossil record.

Lieberman's experience is far from unique. It is the norm. His fossils came from an ecological setting—the Middle Devonian Hamilton Group of eastern North America—in which some 300 species have been discovered. The vast majority of those species are present at the beginning of Hamilton times. Most make it to the very end—when environmental conditions changed and most species became extinct. Many show some signs of change as the six million years went slowly by. But the last members of nearly all those species ended up looking very much like their forerunners six million years earlier. It is a stunning fact that it is possible to illustrate the variation within each of these species in the same small number of photographs or drawings used to depict variation (males, females, juveniles; seasonal plumage changes; geographic variation) in a typical modern guidebook to today's birds in eastern North America. This is stunning because it means that six million years of evolutionary history rarely increased the variation much beyond what was already present within a species over its entire geographic range at any one point in time. This is not what Darwinians of any stripe would expect.

The great stasis debate erupted in the pages of *Nature* in 1987. Once again, John Maynard Smith led the ultra-Darwinian charge. In the very same pages where he welcomed paleontologists back to the High Table of evolutionary discourse in 1984, Maynard Smith wrote a commentary on a paleontological study reported in the same issue. Paleontologist Peter Sheldon had studied the evolutionary histories of eight lineages of Welsh trilobites over three million years. (Trilobites are extinct arthropods—distant relatives of horseshoe crabs, true crustaceans, and insects.) Sheldon examined over 15,000 specimens—an exhaustive study. He claimed that his data support an interpretation of gradual evolution—a result Maynard Smith cheerfully accepted, trumpeted, and brandished as a sword to smite what he saw as the paleontological excesses at the High Table.

Gould and I parried; Maynard Smith thrust again, and we ended that particular flurry with a final riposte. In all, the exchange managed to cover nearly every outstanding bone of contention dividing ultra-Darwinians from naturalists—albeit superficially. We were at blatant cross-purposes, and misunderstanding and distrust of motives is palpable in the exchange (there is a lot of "did not . . . did so" rhetoric in the final exchange). But before the debate went sliding off into peripheral matters, it did manage to highlight the essence of stasis—and the confusion that has surrounded it since we resurrected stasis from the dustbin of nineteenth-century paleontological annals.

To his credit, Maynard Smith did not claim that Sheldon had demolished the existence of stasis as a common evolutionary pattern—at least in his initial commentary accompanying Sheldon's paper. He made the far milder claim that Sheldon's trilobites show that gradual adaptive change does indeed occur. But he saw Sheldon's study as a welcome refutation of Gould and Lewontin's supposed adaptation bashing, and used the opportunity to take a shot at the much misunderstood issue of species selection. Not for nothing was Maynard Smith's piece entitled "Darwinism stays unpunctured."

Almost forgotten, at least by the end of the exchange, was Sheldon's study itself. If nothing else, the debate revealed that one paleontologist's stasis could easily be another's gradual evolution. Sheldon had performed the very useful service of focusing even more attention on the nature of historical evolutionary patterns. That Gould and I looked at Sheldon's results and saw stasis where he saw gradual change served to reinforce the central importance of the actual results of the evolutionary process to debates on how that process operates in nature.

Peter Sheldon has kindly granted me permission to reproduce one of the key diagrams from his *Nature* paper (see next page). Sheldon has conveniently sketched a typical member of each of the eight separate trilobite lineages he studied. Notice that the head—the solid, predominantly white area on the upper portion of each sketch—typically bears a pair of eyes (most easily seen here on *Ogygiocarella* and *Nobiliasaphus*—the two trilobites on the right-hand side of the diagram). (*Bergamia*, *Cnemidopyge*, and *Whittardolithus* are closely related, belonging to a stock of secondarily blind trilobites.) The central portion of a

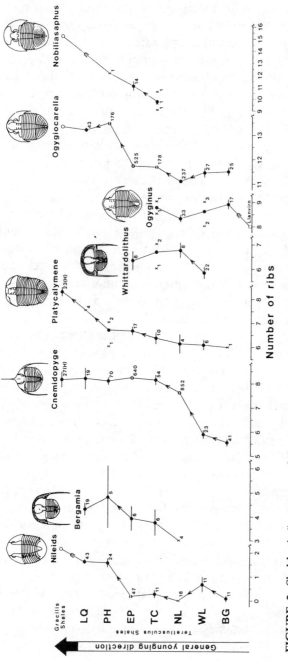

FIGURE 3 Sheldon's "summary of changes in mean number of ribs for eight trilobite lineages in the Builth inlier. Means with 95 percent confidence interval and number of measurements are shown by •ₙ and ●ₙ. Approximate mean (O) based on Hughes [that is, trilobite monographs] or personal observations. Data from Hughes (H). Individual measurements (xₙ). Successive means (▼) are significantly different at the 95 percent confidence level. Successive means (■) are certain to be significantly different but full data are unavailable. Vertical spacing between sections is not to scale. BG: Bach-y-graig Stream Section. EP: Stream Section. LQ: Pencerrig Lake Quarry." From Sheldon, 1987, p. 562, Fig. 4. Reprinted by Permission from *Nature*, vol. 330, 560–563. Copyright 1987, MacMillan Magazines Ltd.

trilobite's body consists of a series of attached segments that allow the animal to roll up—just as a lobster's tail or garden pill bugs (land-living crustaceans) can roll up.

But it was the last portion of the trilobite body—the tail, or pygidium—that provided the focus of Sheldon's study. The tail consists of a series of segments much like the middle part of the body. The difference is that the tail, like the head, is a single, fused piece: its segments cannot flex and roll up. On most trilobites you can still see the segments even though they are fused together.

Therein lies Sheldon's story. He counted the number of segments visible on the tails of specimens from each lineage at different points in the geological sequence (shown by the far left column of his diagram). He found that the average number of segments (which he called "ribs"—referring to the distinct segments on the side parts of the tails) tended to vary over three million years.

Note that the tail of the trilobite furthest left on the diagram (Nileids) shows only a single pair of ribs, the rest of the tail is smooth. Many groups of trilobites appear to have lost conspicuous segmentation of the tail, often as a reflection of adopting burrowing habits. A smooth tail, in other words, is often interpreted as a burrowing adaptation in trilobites. But lack of ribs does not mean a trilobite's tail divisions are lost; rather, they are just not expressed on the tail's outer surface. (Horseshoe crabs are the same way. Hold a juvenile horseshoe crab's translucent rear body portion up to the light, and you can see distinct traces of segments on the inner surface, even though the upper surface is, like the Nileid trilobite in Sheldon's drawing, perfectly smooth.)

In most trilobite species, some individuals have more ribs than others—meaning that some express the tail's segmentation on the upper surface to a greater degree than others. (I have long thought that this variation was basically physiological, relating to the thickness of mineral elements deposited in the animal's shell.) Whatever the basis of the variation, Sheldon clearly showed that the average number of ribs increased independently in a number of separate trilobite lineages. Here—so Sheldon and Maynard Smith pronounced—is a clear-cut case of gradual evolutionary change.

Well, yes and no. Gould and I were quick to point out that at least one lineage (*Whittardolithus*) seems to zigzag and end up where it started. *Ogyginus* zigs and zags between eight and nine ribs. Then, too, three other lineages (Nileids, *Cnemidopyge* and *Ogygiocarella*) seem to increase rib counts, but do so in suspicious ways. In each, the rib count remains remarkably steady (through the point marked "EP" on the time scale to the left of the diagram in two of them; to "WL" for *Cnemidopyge*). Thereafter, the number shifts markedly upwards and then stabilizes. Note that there are no intermediates sampled: Sheldon simply drew a line connecting the two disjunct, stable portions of the lineage. As has become abundantly clear, many sudden anatomical shifts in the fossil record reflect not evolution but migration from elsewhere of related, but different, stocks. We will encounter this pattern again.

Certainly, however, *Platycalymene* and *Nobiliasaphus* show a gradual, more-or-less unidirectional trend to increase in rib count (though one could quibble with these last examples: *Platycalymene* remains quite stable for most of its history, adding an average of two ribs only towards the end of the span of time Sheldon sampled; and *Nobiliasaphus* is well sampled only at a single point in time).

Quibbles aside, Sheldon's trilobites have something important to say about the nature of the evolutionary process. No matter how much gradual change in ribbiness one wants to see in Sheldon's graphs, the real point is that not much happens to any of the eight species lineages in a full three million years of time.

Such minor tinkering with rib number—much of it leading in no particular cumulative direction—can hardly account for the much more substantial anatomical differences between closely related stocks (such as *Bergamia* and *Whittardolithus*, and between *Ogyginus*, *Ogygiocarella*, and *Nobiliasaphus* in Sheldon's own example). This is just the point that George Simpson tried, largely in vain, to convey to evolutionary biologists in the 1940s. Recall Simpson's invocation of whale and bat evolution: the earliest bats and whales, each roughly 55 million years old, are primitive vis à vis modern species, but they are recognizably bats and whales. If one extrapolates the rate of evolution that transformed those primitive Eocene bats and whales to their modern analogues back to the point where each stock diverged from primitive mammals, the results are absurd. Whales and bats would have to have diverged from primitive terrestrial mammals long before any placental mammals had evolved!

Species stasis—the observation that, once they appear, species tend not to accumulate much anatomical change throughout the remainder of their existence—applies Simpson's point nearer to the heart of the evolutionary process. When we look at evolutionary patterns within species, we enter the realm of populations, the locus of natural selection. And what we see, once again, is tinkering—minor oscillations—with some species accumulating minor amounts of change and others zigzagging and waffling around. Even if every zig and zag represents natural selection fine-tuning adaptations to match environmental change, we cannot simply extrapolate such minor bits of transformation and claim we have gotten an accurate, general picture of how natural selection actually works to produce adaptive change in evolutionary history.

Sheldon was far from the first paleontologist to take issue with our claim of the ubiquity of stasis as utterly typical of the history of species. Philip Gingerich, a Yale Ph.D. and thereafter on the faculty at the University of Michigan, quickly entered the fray in the mid-1970s, publishing a series of important papers tracing evolutionary patterns in various lineages of Eocene mammals. Vertebrates, especially terrestrial species, tend to fossilize far less readily than marine invertebrates. But mammalian teeth are something of an exception, and Gingerich was able to amass an impressive collection of meticulously documented specimens through a thick sequence of sediments in the Bighorn Basin of Wyoming. Gingerich measured the surface area of a molar tooth in several different lineages and found that the average size of the tooth tended to change as one compared specimens from different geological layers.

Like Sheldon (and Maynard Smith), Gingerich pronounced his results the vindication of the original Darwinian position of phyletic gradualism. And, just as we did with Sheldon's trilobites, Gould and I disputed Gingerich's findings, seeing in them, instead, predominantly oscillations. We made the point that, all told, not much happened to those mammals over the roughly four-million-year interval Gingerich had sampled.

But it was a population geneticist—Russell Lande, of the University of Chicago—who said the most interesting things about Gingerich's evolutionary patterns. To his great credit, Lande has taken George Simpson's pronouncements on evolutionary pattern seriously. Specifically, he has applied mathematical modeling techniques to probe the nature of rapid evolutionary change of the sort Simpson outlined in his theory of "quantum evolution." But Lande's first foray into paleontological data was a statistical analysis of Gingerich's Eocene mammalian teeth, and what he found is profoundly fascinating.

Lande accepted Gingerich's basic data, which he analyzed using the basic formulae of population genetics. According to Lande, the rate of change in Gingerich's fossil mammal teeth is characteristically so slow that natural selection—if that was indeed what underlay the changes in tooth size—must have been incredibly weak. Indeed, Lande could not rule out the possibility that the changes in Gingerich's mammals were caused, not by natural selection tracking environmental change, but instead by Sewall Wright's genetic drift. Such changes as Gingerich could document could have derived simply from a random sampling of genetic variation from generation to generation through time.

Ultra-Darwinians and naturalists both justly lay claim to the original Darwinian pedigree. All Darwinians affirm that natural selection underlies adaptive change, and I am certainly no exception. But all descriptions of natural selection are like a car in a garage with its engine idling. How that motor actually works—out on the road in terms of Simpson's internal combustion engine—is clearly something else again. If we cannot divine the workings of the internal combustion engine by standing on a street corner watching the cars whiz by, neither can we intuit how natural selection works to effect adaptive evolutionary change by peering under the hood as the engine idles in the garage.

Evolution does not inevitably and irrevocably transform species as they persist through geological time. To the contrary, species most often seem to go nowhere, evolutionarily speaking. To be sure, some will accrue some evolutionary change over millions of years, but most of them hardly accrue any change at all. Stasis is now abundantly well documented as the preeminent paleontological pattern in the evolutionary history of species. Paleontologist Steven Stanley has labored mightily to establish this point through the 1970s and 1980s, publishing original analyses of marine clam evolution. He emphasizes with great clarity and eloquence the most important point that whatever might underlie the weak and typically vacillating anatomical changes that do show up in the fossil record, the little progressive within-species change we see in the fossil record is simply too slow to account for the great adaptive changes wrought by evolution.

Stasis does not mean that we need an alternative to natural selection. But we do need to understand the circumstances in which natural selection works. On the one hand, how does natural selection keep species stable for long periods of time? And what circumstances obtain when natural selection does effect adaptive change? It's a critical point. Much of the food-fighting around the High Table reflects Maynard Smith's contention (well displayed in the Great Stasis Debate in *Nature* that began with his exegesis of Sheldon's trilobite study) that I and other paleontologists seated at the High Table have insisted that something other than natural selection must be invoked to explain adaptive change in evolution.

But such has never been our claim. We are merely dissatisfied with the lack of any cogent theory to explain why natural selection keeps species stable for so long—and what enables selection to trigger change when it does occur. The fundamental question in evolution is not how adaptive change occurs (it comes through natural selection), but why adaptive evolutionary change occurs when it does. Understanding why species tend to remain so constant—even in the face of significant environmental change—is an absolute prerequisite to understanding the circumstances in which the vast bulk of adaptive change occurs in evolutionary history.

What Causes Stasis?

George Simpson captured Darwin's original victory perfectly when he exclaimed that natural selection is such a powerful force that it is impossible to imagine how evolution cannot occur. Darwin replaced "evolution is impossible" with "evolution has happened"—to the permanent satisfaction of the rational world.

We naturalists have refocused the picture. Yes, we say, evolution has happened. And, we agree, in the main, evolutionary history is largely a picture of adaptive change that comes through natural selection. But, we add, evolutionary history affirms that, far from being inevitable, adaptive change comes not with the mere passage of time, but seems to be concentrated at particular times and places with the histories of all lineages. Adaptive change more often than not seems to come as discrete events in evolutionary history. If we can understand why species tend to remain so adaptively conservative throughout the vast bulk of their history, perhaps we can shed some light on those very circumstances that seem to allow natural selection to do what we all agree it eventually does do: produce adaptive change in the history of life.

We have already met one key ingredient to understanding stasis: habitat tracking. Traditional Darwinism assumes that adaptive change tracks changing environments, as natural selection modifies features to keep pace with changing conditions. But, as we have seen, species tend to change locale—rather than anatomical features—in response to environmental change. As long as suitable habitat can be found, a species will move rather than stay put and adapt to new environmental regimes.

Darwinians, going back to the grand old man himself, have always acknowledged that evolution proceeds at different rates at different times during the history of a species lineage. Evolution is generally imagined to proceed faster when conditions are changing and more slowly if the environment is relatively stable. Natural selection is the mediator of evolutionary change, whether fast or slow. When conditions change, directional natural selection modifies adaptations in response. When conditions are stable, species will change little if at all—a mode called "stabilizing selection." When times are stable, so goes the Darwinian canon, evolution sets out to perfect, rather than to modify, adaptation.

Stabilizing selection under such circumstances makes perfect sense. We naturalists differ from standard Darwinian expectations, though, when we see environmental change as a medium of stabilizing selection. If populations within a species move as they recognize the appearance of suitable habitat within reach elsewhere, we would expect stabilizing, rather than directional, natural selection. Thus a profound difference emerges between naturalists and traditional Darwinian thinkers. Both camps agree that stable conditions yield little or no adaptive change. But naturalists also see stasis as the most common outcome of environmental change as well: stasis from stabilizing selection even in the face of environmental change. So long as suitable habitats can be recognized and occupied, there is every reason to suppose that stabilizing selection will continue to operate, or at least to predominate.

But habitat tracking itself does not preclude adaptive change: a species might survive, but it need not do so completely unchanged. My colleague Bruce Lieberman's study of two species lineages of Devonian fossils sheds some light on the relation between habitat tracking, species survival, and stasis. Lieberman worked on two species of brachiopods, bivalved invertebrates that superficially resemble true clams. In both brachiopod lineages, Lieberman found that after six-million-years, each species ended up pretty much as it began. But Lieberman also found that sometime near the middle of that six million year interval, some significant anatomical deviation occurred in both species— a distinct "dog-leg," the zig of a single zigzag evolutionary history.

So far, Lieberman's brachiopods don't seem to have behaved significantly differently in kind than Sheldon's trilobites, Gingerich's mammals, or most other fossil species. But Lieberman was able to delve further. He compared variation between samples, not only over geologic time, but also between samples of brachiopods that lived at about the same time but in different ecological settings. The Devonian rocks that produced Lieberman's brachiopods have many other species of marine invertebrates as well. Most of them persist through the entire interval. But by no means did they all live together in the same ecological setting. Rochester University's Carlton Brett, and his colleague Gordon Baird, have produced a detailed, accurate account of the ten or so distinct communities in which the 300 or so Middle Devonian species of invertebrates existed in what is now the eastern United States.

Brett and Baird have documented a remarkable consistency in the composition of these paleocommunities over the six-million-year interval. Eastern North America lay astride the equator 380 million years ago. The most significant environmental change that we can detect over those six million years was a change in the depth of the seawater—in concert with the waxings and wanings of large-scale deltas that shifted the ancient shoreline around. What was relatively deep and quiet water miles offshore would become, in the passage of Middle Devonian time, near-shore, shallow water lagoons or intertidal mud flats. Each environment supported its different ecological community (though some species were parts of more than one community). As time went by, the shoreline shifted, and the ecological communities followed suit—an example of habitat tracking on a grand scale as entire communities shifted to track preferred environmental conditions.

Here is the twist to Lieberman's study: Both of his brachiopod species occurred in more than one ecological community. Gradual evolution within each species was most pronounced within populations living in nearly identical environmental settings. Peter Sheldon has made a similar point. Typecast as an ardent gradualist by the splash made by his *Nature* paper and Maynard Smith's commentary, Sheldon actually acknowledges that stasis may indeed predominate in the fossil record. He has suggested that habitats with lots of physical environmental change on geological time scales actually engender stasis, either through habitat tracking or selection for ecological flexibility (itself a theme explored in depth in later chapters). More stable regimes (themselves harder to find in the fossil record), on the other hand, provide the context for the accumulation of gradual adaptive change. Sheldon's suggestion neatly turns the traditional expectation of evolution tracking environmental change—and of relative evolutionary stability in the face of environmental stability—on its head.

If habitat tracking were absolutely all that there was to stasis, one would not predict that populations living under the most unchanging conditions would be the very ones that express the most change. Something more must underlie the stasis phenomenon, something beyond simple habitat tracking. Something does: Stasis is an outcome of the organization of species in the wild. And our appreciation of that structure and its evolutionary implications comes from the work of a population geneticist: Sewall Wright. The neo-Darwinian failure to grasp the implications of Wright's work in the 1930s, plus the competing notions recently espoused by ultra-Darwinians to explain stasis, tell us much about the reductionism of traditional evolutionary biology. And it reveals vividly the lack of importance ultra-Darwinians attach to the organization of living systems in the wild.

Sewall Wright: Populations, Species, and Stasis

Shortly before he died, Sewall Wright rather testily complained that he had anticipated the punctuated equilibria of Eldredge and Gould by a whopping 40 years. That just added one more to the list, as Simpson and Mayr had already made similar claims. Each claim had a kernel of truth. We drew from Simpson an appreciation of historical pattern, particularly his realization that evolution comes in brief spurts "punctuating" vastly longer periods of much slower rates of evolution. Our difference with

Simpson was that he still subscribed to the Darwinian tradition of phyletic change for species. Simpson's ideas on quantum evolution were geared for bigger game: the origin of large-scale groups, such as mammals and dinosaurs. But we saw the pattern at the level of species themselves—far closer to the traditional locus of evolutionary action according to the Darwinian canon.

We owed Mayr and Dobzhansky a different debt: an appreciation of the importance of really understanding the exact nature and mode of the evolution of species. What they missed, of course, is the fundamentally static nature of species' histories once they evolve. That's where Sewall Wright comes in—not from any appreciation of the facts of the matter, but from a purely theoretical point of view. Stasis is implicit from Sewall Wright's treatment of the organization of species in nature.

By all accounts, Sewall Wright was no field man. He started out professional life working on the genetics of guinea pigs and analyzing data from livestock breeders. By the 1930s Wright had turned his attention to evolution. He was embroiled in a lengthy wrangle with England's Ronald Fisher, who thought of evolution solely as selection-mediated change within populations— the basic theoretical stance forming the core of ultra-Darwinism to this very day. Wright, recall, had his own view of things, embodied in his "shifting balance theory," which allocated a substantial role in the evolutionary history of species to chance ("genetic drift").

Essential to Wright's theory was his view of the nature and structure of species. It is remarkable that Dobzhansky, who was the preeminent field man (at least among North American geneticists), adopted Wright's views on the organization of species. Wright, relying perhaps on no more than a commonsense appreciation of the distribution of organisms (though undoubtedly also influenced by fellow University of Chicago ecologist W. C. Allee and, later, evolutionary biologist and termite specialist A. E. Emerson), gave us the fundamental view of species organization still with us today: species are composed of a series of semi-isolated populations.

Look in any field guide, and the ranges of most species will be encompassed by a single irregular line. Most species (especially in the higher latitudes) are distributed over large regions, often a half continent and sometimes more. There are invariably a multiplicity of different habitats within such large areas, and no species will be found in all habitats. The result: Species are inherently and necessarily disjunct in their distributions, despite the nice neat line that can be drawn around their entire range of distribution.

It is this disjunction, which greatly reduces the opportunity for genes to be exchanged between populations, that inspired Wright to envision the semi-isolated populations within a given species undergoing semi-independent evolutionary histories. In modern parlance, local populations of a species will be integrated into local ecosystems, and each ecosystem will differ somewhat from the next. This is especially true of species adapted to a relatively broad spectrum of habitats. But it is true to some extent of all species, no matter how narrow the range of habitats their organisms can exploit.

Now, imagine a species persisting as geological time rolls by. Each population represents a sampling of the genetic information in the species. Each is in different ecological circumstances. The effects of drift and especially natural selection will be different in each semi-isolated population. In addition, some populations will become extinct, while two adjacent populations may fuse. As time goes on and environmental conditions change, the effects will be different from one population to the next.

Given this picture of the organization of species, it defies credulity that any single species, as a whole, will undergo massive, across-the-board gradual change in any one particular direction. After all, its component populations are all having semi-independent histories, some going one way, some another, others still yet another. There is no way the entire species can march off, all demes united, on a single evolutionary pathway. The evolutionary result for the entire species lineage might best be predicted as no net change at all. Stasis is fully, and obviously, implicit in Wright's view of the structural organization of species. I believe this is what Wright had in mind when he said in

1982 that he had already arrived at a concept of "punctuated equilibria" in the 1930s.

And there is a tantalizing corollary. If Wright's image of species organization predicts no gradual, cumulative evolutionary change for an entire species, it also predicts that gradual change may well show up in the protracted histories of parts of species. Evolutionary histories within relatively localized, disjunct populations may well show some directional, gradual change—even if the species as a whole exhibits little or no net change throughout its entire existence. In fact, that's just precisely what Bruce Lieberman found in his two species lineages of brachiopods: phyletic change was greatest within populations of a particular ecological community type.

Then, too, there are the two most celebrated examples of phyletic evolution brought forward in the stasis debate over the past 20 years: Phil Gingerich's Eocene mammal teeth, and Peter Sheldon's trilobite tails. Gingerich restricted his initial studies to Wyoming's Bighorn Basin. The species he sampled, however, were present in other such semi-isolated basins throughout the west. Paleontologist David M. Schankler has shown that the evolutionary patterns within at least one of Gingerich's lineages were heavily influenced by migration in and out of the Bighorn Basin. Indeed, Schankler interprets the bulk of supposed evolutionary change within the *Phenacodus* lineage as due to shifting distributions of entirely separate *species*—each one remaining in relative stasis while present in the Bighorn Basin.

Much the same may to be true for Sheldon's trilobites. Collected from a restricted region in Wales (called the Builth Inlier), his eight species lineages undoubtedly occurred elsewhere as well. In this case, no one has yet checked to see what sorts of evolutionary patterns are to be found in other populations of these lineages living outside Sheldon's collecting area. Wright's work tells us that stasis should be the overwhelming expectation for entire species, while some degree of phyletic change is to be expected within more localized subdivisions of a species. That appears to be exactly what we do see in well-preserved and well-studied fossil sequences.

It is astounding, in retrospect, that no evolutionary geneticist saw the implication of Wright's view of species organization. Stasis is there, staring all of us in the face. It is there in Wright's basic characterization of species organization—a view warmly embraced by Dobzhansky and accepted by all field biologists. Ultra-Darwinians accept the view as much as any other group, but have ignored its implications at least insofar as predicting the larger patterns of evolutionary change. And stasis is there, as well, staring out at us from the fossil record. It's high time we took stasis seriously.

Ultra-Darwinians Confront Stasis

Although Maynard Smith tweaked us on the ubiquity of stasis when writing his commentary on Sheldon's study in *Nature*, his basic stance has been that stasis seems a real phenomenon—one that begs explanation. Of course, as he and other ultra-Darwinians are quick to add, what we paleontologists have to say about the causes of stasis is wrong. Only evolutionary geneticists, privy as they are to the workings of the evolutionary motor, are in a position to explain stasis.

Maynard Smith bolsters that particular claim by saying that paleontologists rely on some sort of metaphysical belief in "species homeostasis," whereas geneticists are more inclined to ascribe stasis to natural selection. This is remarkable. Gould and I alluded to homeostasis at the end of our original paper in 1972. Since then, we, along with our paleontological confreres, have always emphasized that it is habitat tracking—stabilizing natural selection in the face of environmental change—that basically underlies stasis.

Ultra-Darwinians focus on natural selection within populations. In that, of course, they are conservatives, deeply imbued with the Darwinian tradition. But to seek to reduce all of evolutionary history—even the history of adaptive stasis and change—to the dynamics of population genetics, ultra-Darwinians must argue that species are not special kinds of entities. Indeed, ultra-Darwinian rhetoric frequently and freely interchanges the words "species" and "populations." In so doing, of course, they miss the very distinctions that Wright was at pains to establish. They also depart from the critical work of Ernst Mayr and Theodosius Dobzhansky, who addressed evolutionary discontinuity in the context of the nature and evolution of species, a theme that we naturalists have picked up and amplified in recent years. As we shall see, adaptive change is very much bound up with the process of species formation. But to an ultra-Darwinian, species really do not exist—at least not in any sense distinguishable from "populations." Ultra-Darwinians are really followers of Ronald Fisher—seeking to explain all evolutionary phenomena strictly in terms of natural selection acting on heritable variation within populations.

Nowhere is this clearer than in the work of George Williams, whose *Adaptation and Natural Selection* (1966) is the closest thing to a founding document of post-1959 ultra-Darwinism. His more recent *Natural Selection: Domains, Levels and Challenges* (1992) pursues his exploration of adaptation, and takes up a number of issues that have surfaced over the intervening years. He is especially and explicitly critical of those of us who see anything special about the nature and organization of species.

Williams spends a brief chapter on stasis in this recent book. With Maynard Smith, Williams accepts stasis as a phenomenon meriting serious attention. As he puts it, though lay people have often wondered if there has been enough geological time for evolution to produce the great diversity of living form we see around us, the real question is why have organisms "not done nearly as much evolving as we should reasonably expect." Williams is thoroughly imbued with the workings of the purring motor of evolutionary change, seeing variation and adaptive change everywhere. Like Simpson before him, it is impossible for Williams to imagine evolution not happening—and this means everywhere, all the time.

As a cause of stasis, Williams dismisses habitat tracking as "fable." He offers no countervailing evidence, and seems to make nothing of the migration of species in response to climatic change. After all (Williams says), ecological communities are invariably reorganized: the species represented in an arctic tundra community will differ in detail from those found in a tundra community established around present-day New York at the height of glacial advance. Selection is bound to be somewhat different (we all agree) in such slightly different circumstances—and must produce significant evolutionary change. Once again, we have the a priori extrapolationist view—unsullied by any reference to examples in nature.

Evolution is as much about the fate of adaptive change as it is about its generation. Wright, too, saw evolutionary divergence between his semi-isolated populations as inevitable—a product not only of different selective regimes, but also of mutational history, the initial sampling of genetic variation, and other factors. Wright taught us (and Darwin also saw) that what goes on in an evolutionary sense within a species is for the most part lost. Local populations become extinct, or meld with others over the course of time, so that much of the evolutionary change within a species is simply lost in the long-term course of events. As we shall see, the difference between a species and a population is absolutely critical. Species represent a level of permanence that acts to conserve adaptive change far beyond the ephemeral capacities of local populations.

Williams also sees the fate of adaptive change as central to an understanding of stasis. Because evolutionary change is so rampant, so inevitable to Williams, stasis must be caused by the selective removal of the adaptive changes that do occur. This notion Williams aptly labels "a desperation hypothesis."

Here Williams does refer to a case drawn from nature: the evolution of three-spine stickleback fish as analyzed by paleontologist Michael Bell. Williams sees a persistent ancestral marine species repeatedly giving rise to freshwater descendants—which Williams variably calls "forms" or "populations." They are, in all likelihood, actually separate species (a point Williams conceded in a recent conversation). Whatever their actual status might be, because of their inability to compete successfully, or perhaps because their habitats are prone to disappearance, these freshwater "forms" are continually weeded out, leaving only the ancestral species. Thus stasis, to Williams, is not the absence of evolution, but the systematic loss of evolutionary innovation through a differential extinction process that he calls "normalizing clade selection."

Elsewhere in his book, Williams heaps the kind of scorn common among ultra-Darwinians on the idea of "species selection"—a notion emanating from naturalists in conjunction with discussions of punctuated equilibria. *Species selection* refers to the differential rates of appearance and extinction of species within lineages, and has proven useful (as we shall see) in the explanation of certain large-scale patterns, such as evolutionary trends. It is remarkable that to produce an explanation for stasis Williams had to invent an elaborate mechanism very like "species selection" to account for a phenomenon that I and other naturalists see as perfectly explicable in basic Darwinian terms. We simply see that natural selection acts to stabilize a species as its organisms continue to locate recognizable habitat. Habitat tracking, combined with an appreciation of species structure, developed first by Sewall Wright, is really all that is needed to explain stasis.

But the truly odd thing about Williams's discussion of stasis is that he hardly discusses the persistence—unchanged—of the ancestral species itself. What he does say seems to belie his rejection of habitat tracking. His two-paragraph discussion of the subject concludes that "the appearance of stasis in the fossil record would result from an enormous variability in the persistence of ecological niches." Needless to say, it is the "niche" of the "ancestral type" that, according to Williams, is often the one that persists. I am hard-pressed to see any meaningful distinction between Williams's casual invocation of relative persistence of niche types and the notion of habitat tracking.

Williams is not the only ultra-Darwinian to tackle the issue of stasis. John Maynard Smith has also done so, to rather better effect than Williams. But Maynard Smith's encounter with stasis is also deeply marred by the same theoretical stance adopted by Williams. Maynard Smith also misses the difference between populations and species, as is all too evident in his application (with Swedish biological colleague Nils Stenseth) of the Red Queen hypothesis to the origin of evolutionary pattern.

The Red Queen: Personification of Gradualism

Leigh Van Valen is an imaginative evolutionary biologist who has spent his entire professional career at the University of Chicago. He was trained in New York under the joint aegis of paleontologist George Gaylord Simpson and geneticist Theodosius Dobzhansky. The results have been interesting. They include a modern incarnation of gradualism, as well as a further ultra-Darwinian grappling with the phenomenon of stasis.

In 1973, Van Valen proposed his "Law of Constant Extinction." In group after group, so Van Valen claimed, the evidence is that species undergo a characteristic and fairly steady rate of extinction. This is in itself an interesting proposition—one that has not gone unchallenged by a disparate group of critics. Whether or not Van Valen's contention is true, it is the explanation of the phenomenon that distills the essence of ultra-Darwinian thinking on adaptive evolutionary change.

What could possibly be causing constant extinction rates? Van Valen once remarked that evolution is the outcome of an interplay between ecology and developmental biology. It is to his credit that Van Valen sought an evolutionary explanation of constant extinction rates in an explicitly ecological context.

Van Valen imagined a world in which the evolutionary fate of a species depends very much on what is happening simultaneously to all the other species living in the same ecological setting. To Van Valen, it seemed obvious that the least change affecting any one species would be sure to affect other species in the same ecological system. If, for example, two species each partially depend on a particular food item, and one of them evolves a greater efficiency in exploiting that resource, it makes sense to assume that the other species will be adversely affected.

Van Valen simply generalized this picture, asserting that any and all evolutionary changes within species in an ecological system would have negative consequences for one or more other species in the system. It is a "zero sum" game in which all evolutionary change—understood, of course, as improvements—has to be counterbalanced by negative effects on other species. These negative effects become evolutionary spurs, goading the negatively impacted species to evolve in response to the challenge.

A stark vision of nature indeed. To Van Valen, even if the physical environment is not changing, neighboring species are, and that spells, in effect, environmental degradation for each species. Just like the Red Queen in *Through the Looking Glass*, each species is forced to keep running just to stay in the same place—to keep changing to adjust to the ever changing, ever deteriorating conditions of existence. Species are imagined to be under constant pressure to change, with natural selection hard at work in each generation to tighten each species' ever-precarious grip on existence. Van Valen's Red Queen hypothesis is the modern epitome of traditional Darwinian gradualism, an imagery and theoretical stance that has been used extensively by ultra-Darwinians. Elisabeth Vrba got it dead right when she called the Red Queen the very personification of gradualism.

How reasonable an expectation is Van Valen's "zero-sum game"? Ecologists have long debated the degree to which various species present in a local ecosystem are mutually dependent. Van Valen assumed a great deal of mutual interaction and interdependence among species within ecosystems when he formulated his Red Queen hypothesis. Such systems are said to be "highly accommodated," and biologists have long known that some ecosystems are more highly accommodated than others. What happens to the Red Queen in ecosystems that are not so tightly wrapped—where species are far less mutually interdependent than Van Valen assumed to be the universal case?

John Maynard Smith, working with evolutionary biologist Nils Stenseth, asked precisely this question, as the two set out to explore further ramifications of the Red Queen. Their approach was purely theoretical, building on the mathematical approaches of population genetics initiated by Sewall Wright, Ronald Fisher, and Maynard Smith's mentor, J. B. S. Haldane. They found that Van Valen was absolutely right. In highly accommodated ecosystems, gradual evolutionary change was the general outcome for all species in the system.

But what of species in ecosystems with relatively little accommodation? Their calculations showed that, in such circumstances, evolution seems to grind to a halt. Finally, a decade after we paleontologists had brought the phenomenon of evolutionary stasis back into the arena of evolutionary analysis and discussion, ultra-Darwinians had found stasis staring at them in their own results.

Or had they? Though Maynard Smith has repeatedly acknowledged the importance of stasis, he and Stenseth insisted that the sort of stasis latent in their results was somehow different from the patterns of stasis that I and a number of my paleontological brethren were increasingly pointing to in the fossil record. The reason: They had, they claimed, uncovered a causal explanation for stasis consistent with the principles of neo-Darwinian evolutionary mechanics. This was something paleontologists, they felt, had not managed to achieve.

But Stenseth and Maynard Smith had not succeeded in explaining stasis of species. Their study was designed to study the fates of local populations of species integrated into ecosystems, rather than the evolutionary history of an entire species lineage, as Yale paleontologist Elisabeth Vrba was quick to point out.

Once again, ultra-Darwinians had blurred the distinction between population and species simply to make the paradigms of population genetics fit patterns of evolutionary history. Sewall Wright's careful dissection of species structure—and the complex genetic fate of species over time—was (as usual) completely ignored in their argument, which is regrettable. Stenseth and Maynard Smith had actually made a contribution to understanding why gradual change might show up in the course of time within some populations within a species, and why some populations might remain relatively stable. But they failed to explain stasis as a species-wide phenomenon simply because they did not really address it.

Adaptive change—its origin, maintenance, and further transformation—is, of course, the heart and soul of evolution.

But the ultra-Darwinian reliance on a 150-year tradition of extrapolating natural selection through geologic time to yield the adaptive diversity of life simply doesn't work. It flies in the face of the facts of history—the sorts of evolutionary patterns that Simpson, in his younger days, was so eager to see incorporated into the mainstream of evolutionary discourse. It is these patterns of stasis and change that provide the point of departure for our naturalistic, paleontological contributions to conversations at the evolutionary High Table.

Everyone agrees that natural selection is the deterministic process that takes variability and shapes it into modified organic form. It is the motor of adaptive evolutionary change. But it has become abundantly clear that ultra-Darwinians—focused as they are on genes and populations—are utterly unable to specify the general conditions governing adaptive change. When and to what extent evolutionary change actually occurs in evolutionary history is what I call the context of evolutionary change.

With their heads stuck under the hood while they tinker with the engine, ultra-Darwinians know the principles of what moves the car, but they cannot imagine the road conditions under which it is driven. We naturalists have had our eyes on the road for some time now, and believe we can do a lot better in explaining the context in which selection produces stability as well as adaptive change. And that context hinges critically on the nature and structure of species.

4

Evolution in Real Time

Punctuated Equilibria and the Eternal Species Wrangle

No idea has excited more interest, sparked more debate, been more widely cited, and been more profoundly misunderstood in the post-1959 annals of evolutionary biology than the notion of "punctuated equilibria" that I published with Stephen Jay Gould in 1972. Some of its supporters have touted its revolutionary insights, signaling the end of the Darwinian era. Some of its critics have damned it as, at worst, incredibly wrong-headed; at best, a minor footnote; and, at most, a rip-off of what had already been said before.

There is no question that our publication of "Punctuated Equilibria: an alternative to phyletic gradualism," in an edited volume aimed at raising biological consciousness in the minds of our paleontological brethren, was itself something of a punctuational event. It snapped evolutionary-minded paleontologists back to attention, and within a few years had served notice to the evolutionary community at large that paleontologists had once again joined the fray. Also, it sent the distinct message that we were once again making trouble for Darwinian orthodoxy. It was punctuated equilibria that brought paleontologists back to the

High Table, and served as the centerpiece for most of the heated discussion that has ensued.

Punctuated equilibria itself is a remarkably simple idea. It is a melding, in essence, of the pattern of stasis with the recognition that most evolutionary change seems bound up with the origin of new species—the process of speciation. Stasis upset the applecart in the 1970s, with some paleontologists hotly denying that the Darwinian vision of change through time was refuted by conventional experience with fossil lineages. Others, of course, told us immediately that they had known it all along. Paleontologist Dick Bambach, who has spent his entire career at Virginia Polytechnic Institute, was doing his Ph.D. research at Yale in the late 1960s, while Gould and I, with our fellow graduate students, were wrestling with evolutionary problems at Columbia and The American Museum of Natural History. Bambach later told me that he surveyed over 70 species lineages from Silurian rocks spanning some 30 million years in Nova Scotia. He reported on the three lineages that seemed to show some evolutionary change. He did not dwell on the remaining 70-odd lineages that showed little or no change. But it was experiences such as his that led many paleontologists to tell us that stasis is indeed the rule, not the exception.

Well, if evolutionary change doesn't simply accumulate over the course of time, the question becomes, When and under what conditions does evolutionary change occur? Here we confront a related observation, one that stands almost as a corollary of the very recognition of stasis: new species—for the moment meaning organisms with novel anatomical characteristics—tend to show up abruptly in the fossil record as the overwhelming rule.

But we must be careful. Abrupt appearances of species may mean—and in fact, very often do mean—that the species migrated into the area in response to changing environmental conditions. As we have already seen, species are constantly expanding and contracting their geographic ranges in direct response to environmental change. If we know of a species' prior existence in one place, its abrupt appearance elsewhere is a sure sign of such migration. But one cannot be so sure when dealing with the earliest known appearance of a species whether we are looking at it just as it has evolved or if it came in from someplace else.

No wonder paleontologists shied away from evolution for so long. It seems never to happen. Assiduous collecting up cliff faces yields zigzags, minor oscillations, and the very occasional slight accumulation of change—over millions of years, at a rate too slow to really account for all the prodigious change that has occurred in evolutionary history. When we do see the introduction of evolutionary novelty, it usually shows up with a bang, and often with no firm evidence that the organisms did not evolve elsewhere! Evolution cannot forever be going on someplace else. Yet that's how the fossil record has struck many a forlorn paleontologist looking to learn something about evolution.

From an evolutionary point of view, then, the fossil record has long had two strikes against it: its gappiness, and uncertainties about where its fossilized animals and plants might have come from. Darwin devoted two chapters of the *Origin* to geological time and the fossil record. In the first (chapter 9, "On the Imperfection of the Geological Record"), he established the enormity of geological time, thus providing himself all the time necessary for life to evolve, and making a great contribution to our understanding of earth history. In that same chapter Darwin was at pains to explain why paleontologists had not as yet found numerous examples of what he called "insensibly graded series" of fossils. Why, in other words, had paleontologists not been able to document the very pattern Darwin thought evolution must leave in the rocks?

For one thing, Darwin mused, paleontology in his day was still in its infancy. Surely, he wrote, paleontology would eventually provide full corroboration of his theory. Darwin actually believed that his entire theory of "transmutation" (or "descent with modification"—he never called it "evolution" in the *Origin*) would stand or fall on the eventual recovery of many examples of gradual evolution in the fossil record.

But the collective inexperience of paleontologists did not completely allay Darwin's qualms. So, brilliant and thorough thinker that he was, he essentially invented a new field of scientific inquiry—what is now called "taphonomy"—to explain why the fossil record is so deficient, so full of gaps, that the predicted patterns of gradual change simply do not emerge.

It is absolutely the case, as Darwin was among the first to establish, that only the minutest fraction of all organisms ever make it into the fossil record. Those that do are routinely destroyed by chemical and physical events. Some are buried so deep we never find them, while others erode out and dissolve away before they can be collected. The chance that any particular fossil will ever make it to the laboratory bench is vanishingly small. It's a wonder we can say anything intelligent at all about fossils, and small wonder, Darwin concluded, that the rock record is not more cooperative in supplying abundant examples of gradual evolutionary transformation.

But, we saw—as did several paleontological contemporaries of Darwin—that if you do collect a series of fossils up through a sequence of sedimentary rock, and if you don't see much evidence of anatomical change through that series, that is indeed evidence that substantial gradual evolutionary change has not occurred within that species lineage, no matter how gappy the record may be. That's why the evidence for stasis now appears so overwhelming—and has, as we have seen, gotten the attention of ultra-Darwinian theorists.

I simply thought that the time had come to take the fossil record—the patterns of stability and change—a bit more literally than had traditionally been the case. George Simpson had begun the process when he insisted that gaps do not explain away the abrupt appearances of large-scale taxa—meaning, large-scale events of evolutionary change. Simpson was perfectly content to blame the absence of examples of gradual change within and between species on gaps in the record, but found (to his everlasting credit) that the argument could not be stretched to encompass large-scale evolutionary change, such as the derivation of whales or bats from terrestrial mammalian precursors.

I simply extended Simpson's argument to the level of the species. This is much closer to the nexus of paleontology and genetics, and near, if not precisely at, the traditional locus of the evolutionary process: natural selection within local populations. The persistent pattern of nonchange within samples, coupled with the abrupt appearance of new species—organisms marked with anatomical innovations—had to be telling us something about the way the evolutionary process works. After all, stasis was telling us that the old Darwinian picture couldn't really be entirely right.

But I needed something more than pattern. I needed to explain why evolution leaves an entirely different sort of pattern in the rock record than Darwin—and his long string of successors, including many paleontologists—had supposed. And I found a very ready source of explanation staring me right in the face. I found it in Dobzhansky's and Mayr's work on species and the nature of the speciation process, specifically the derivation of descendant species from ancestral species through geographic isolation. Thus developed the combination of pattern and process that Steve Gould and I called "punctuated equilibria" in our joint 1972 paper that came hard on the heels of my original 1971 analysis: evolutionary pattern (stasis with its corollary of relatively abrupt evolutionary change—stability *punctuated* by change) plus mechanism (natural selection in the context of speciation). Speciation, the fragmentation of an ancestral species into two or more descendants, is a component of the evolutionary process. It takes speciation, it seems, to break the stranglehold of stasis, providing the context for lasting evolutionary change. Punctuated equilibria is simply the notion of speciation applied as the explanation for evolutionary change interrupting vastly longer periods of monotonous stasis. It should have been noncontroversial. It wasn't.

The Early Storms Over Punctuated Equilibria

The Great Stasis Debate that erupted over Peter Sheldon's trilobite study in 1987 was by no means the first challenge to our claim that stasis is the evolutionary norm. Phil Gingerich, while still a graduate student at Yale, was one of several paleontologists quick to jump in with data purporting to support Darwinian orthodoxy against our heretical notion of stasis.

But there were other issues, as well. Gould proposed a follow-up paper (published in 1977) to reply to our early paleontological critics, and to bring the idea of punctuated equilibria to a wider, more expressly evolutionary biological audience. That "son of" paper did bring us closer to the High Table, but also immediately provoked negative reactions. Among the charges, perhaps the most prominent and important one accused us of abandoning the Darwinian camp altogether, promulgating instead a form of saltationism (meaning literally evolution "in jumps"). This charge even emanated from the High Table, most notably from Ernst Mayr—whose reactions to punctuated equilibria have been various and not a little self-contradictory.

Richard Goldschmidt was perhaps the favorite whipping boy of the modern synthesis as it was being developed and promulgated in the early 1940s. We have already seen (chapter 1) how his saltational notions helped unify the neo-Darwinians. All those seated around the High Table back then took their pot shots at Goldschmidt and his notion of "hopeful monsters." And, truth be known, Goldschmidt provided little in the way of plausible genetic mechanisms to bolster his credibility. Gould, I, and all our fellow students were duly exposed to Goldschmidt's ideas, held out as an aberrant, even ludicrous, departure from the neo-Darwinian paradigm. As basic adherents of that paradigm, we were all happy enough to agree.

But we had proposed that evolution seems to come in spurts, only occasionally breaking the monotony of vastly longer periods of virtual evolutionary stagnation. And there really is a problem communicating concepts of time between various biological disciplines. It is easy enough for a paleontologist to grasp the brief intergenerational time frame of geneticists. Population geneticists for the most part deal with generations, which can range from a few hours in microbes to decades in slow-maturing mammals such as ourselves. But geneticists, understandably, might have a bit more difficulty with the thousands, tens of thousands, hundreds of thousands, and even millions of years that constitute the standard paleontological time frame.

Generalizing a bit, Steve and I said that most species of marine invertebrates—that vast bulk of species of the fossil record—last between five and ten million years, a rough estimate that has actually stood up pretty well over the years. Some species last longer, and many become extinct far sooner. Terrestrial animals—of all kinds, including insects, snails, and vertebrates—appear not to last as long as their marine counterparts. Extinction rates are characteristically higher on land. This is testimony to the more immediate and frequent effects of climate change on land when compared to habitats below the waves.

As against five to ten million years of stasis, we claimed that evolutionary change—tied up in speciation events—happens rather quickly. Here we are at the smallest level of resolution of geological time often (but not always) possible with the fossil record. Even tens of thousands of years are usually difficult to decipher in the fossil record. So our estimates of time required for speciation events were much hazier than our estimated average durations of species. I came up with the figure "five to fifty thousand years," which was consistent with some of the events we believed we had some direct data on from our own studies.

The real point, of course, is that "five to fifty thousand years" is in the neighborhood of 1,000 times shorter than the average durations of species. It is the contrast in rates—between vast periods of essentially no change, and brief intervals of actual change—that is most important. What we were saying is that evolution looks instantaneous in the fossil record, but is not. Indeed some evolutionary geneticists have said that the "five to fifty thousand years" estimate is, if anything, overly generous. Speciation events may often require even less time to take place.

Nonetheless, we were accused of being saltationists. Steve Gould wrote two consecutive essays on Richard Goldschmidt in his monthly column in *Natural History* in 1977. Among other things, Steve speculated that the recent discovery of regulatory genes—genes that turn other genes on and off—raised the possibility that mutations in the regulatory apparatus might occasionally have the sort of effect Goldschmidt had in mind with his notion of "macromutations." These macromutations had the large-scale effects of the sort he posited for his "hopeful monsters." Nowhere in either article did Steve mention punctuated equilibria.

But it was enough, it seems, that he, champion of a new model positing bursts of relatively rapid change, would, a few years later, discuss Goldschmidt in favorable terms. Mayr was one of the first to level the charge that punctuated equilibria is nothing but old saltationism in new guise. Our debt to Mayr's concept of species and speciation, so central to the idea of punctuated equilibria, eventually induced him to do an about face. Mayr came to prefer taking credit for punctuated equilibria rather than seeing it linked to his old nemesis Goldschmidt.

Gould and I were regularly derided and dismissed as neo-saltationists for many years thereafter, and still occasionally hear the charge. A highlight, if such it be, was the clever if annoying label attached both to punctuated equilibria and to its parents by British geneticist J. R. G. Turner. His sole contribution to the debate was to characterize punctuated equilibria as "evolution by jerks." And I confess it is an irony that makes me cringe just a bit that my book *Time Frames,* written as an introduction to and history of punctuated equilibria—owes its Italian publication to the Hopeful Monster Press.

Suffice it to say that Russell Lande is right: five to fifty thousand years is right on the money, even a bit of an overestimate, for normal Darwinian within-population processes to effect the average (generally not very great) amount of change that occurs in conjunction with speciation events. Whatever we are, we are not saltationists!

Another contretemps with punctuated equilibria at the center erupted in the late 1970s, consuming several gallons of ink—once again in the pages of *Nature.* This particular food fight took place in another part of the dining hall, a bit away from the High Table per se. It nonetheless shed some light on the overall cultural and social matrix in which the scientific enterprise is embedded. Evolution, more than any other general scientific proposition, is fraught with implications far transcending the narrow confines of biology.

In our 1977 paper, we discussed Darwin's insistence on gradualism, suggesting that he, like any normal human, was most likely affected by the prevailing sentiments of his day. The rise of a prosperous middle class during the Industrial Revolution had much to do with the zeitgeist of Darwin's times and seemed relevant to us. The watchword of the day seems to have been "progress," and indeed, writing a decade after the *Origin* appeared, Benjamin Disraeli said, "Change is inevitable in a progressive country." Change, it seems, was acceptable—so long as it was slow, steady, gradual change for the better.

Now, I hasten to add, I am fully convinced there were other, more potent factors underlying Darwin's dedication to gradualism. There was his debt to Lyell, who established gradualism in geology as part of a more rigorous approach to science generally, in his efforts to combat the catastrophists. There was the anti-evolution that was part and parcel of prevailing religious belief in Great Britain at the time—certainly including Emma Wedgewood Darwin. Gradualism softens the naturalism and materialism—the godlessness—of evolution at least somewhat. And then there is Darwin's apprenticeship as a pigeon fancier. His attribution to Nature of a process akin to artificial selection, and his projection into geological time of the sort of directional change (for the better) that could be wrought by astute animal breeders over a few generations, was certainly both logical and the closest thing possible to empirical, even experimental, verification of his ideas.

Nonetheless, Darwin certainly was influenced by the climate of social and political thinking of his day. Making that point, we added that no one is immune from such influence, noting that "it might also not be irrelevant to our personal preferences that one of us learned his Marxism, literally at his daddy's knee."

Well. As if we had laid our souls bare in *True Confessions*, punctuated equilibria was seized upon as a Marxist tract, plain and simple. Leading the charge was fractious (and now, sadly, deceased) British paleontologist Lionel Beverly Halstead—who,

I am told, was a Marxist himself in his student days. Having seen the light, and knowing a Marxist (especially a supposedly self-confessed one) when he sees one, Halstead wrote to *Nature* with an astounding proposition: Her Majesty's schoolchildren were being subjected to Marxist propaganda as they viewed the newly renovated dinosaur exhibit at the Natural History Museum in South Kensington!

How did Halstead arrive at such a bizarre scenario? It turns out that punctuated equilibria was but one of several relatively new propositions that irked Halstead. Another was cladistics, that new school of systematics that emphasizes genealogy in the reconstruction of evolutionary history (see chapter 2). Halstead hit the roof when he learned that cladistically imbued paleontologists at the (British) Natural History Museum had reorganized the dinosaur exhibits along cladistic lines. Traditionalists are still infuriated when a dinosaur paleontologist tells them that birds are just dinosaurs with feathers.

But how did Halstead turn the cladistically arranged dinosaurs into a Marxist brainwashing? Simplicity itself: Halstead proclaimed cladistics to be based on punctuated equilibria. Punctuated equilibria, in turn, was nothing but a Marxist tract masquerading as a serious scientific proposition!

Now, Gould is no cladist; whereas I quickly adopted cladistic methods in my trilobite research. Ornithologist Joel Cracraft and I wrote one of the earlier books on the subject. I am no Marxist, and neither for that matter is Steve; learning and adoption are two different things. Marx, I understand, also denied being a Marxist. Darwin was probably not a Darwinist, might well have become a neo-Darwinist, and, I feel sure, would reject ultra-Darwinian excess and be a naturalist—as he was all along! My cladist colleagues can only smirk at the suggestion that cladistics derives from punctuated equilibria. Cladistics, recall, has been ridiculed in ultra-Darwinian quarters because of the apparently antievolutionary stance adopted over the last decade by some of its more prominent practitioners.

Punctuated equilibria is no Marxist fantasy. Gould added the remark to explain why discontinuity and abrupt change might appeal to him for reasons beyond scientific evidence. (On this very point, some ardent Marxists have claimed that punctuated equilibria is in fact inconsistent with true Marxism—or, if Marxism, it is very bad Marxism.)

Fair enough. I am still not quite sure what to make of the zeitgeist of our own times, in which, quite apart from Marxism (or so I believe), paradigms similar to punctuated equilibria have shown up in a wide range of academic pursuits. Before us, there was the celebrated case of Thomas Kuhn, whose *Structure of Scientific Revolutions* became a best-seller in the early 1970s. His central thesis was essentially that science proceeds as status quo paradigms in stasis, interrupted by rapid events that finally throw out the old paradigm in favor of a new one that handles all the anomalies swept under the table by its predecessor. Catastrophe theory, similarly, became a hot topic in mathematics as the 1970s wore on. Archeology and political science have also seen new theories emerging along similar lines (some, I am happy to report, explicitly derived from punctuated equilibria). There may well be a general reaction against models stressing smooth, linear continuity emerging in the intellectual community generally—models of which punctuated equilibria is but one example. As we shall see later (chapter 6), the hierarchical view of the biological world that comes directly out of punctuated equilibria dovetails nicely with general work in hierarchy theory and with chaos theory, that very recent and quite pervasive hot topic.

So much takes us rather far away from the High Table. Yet I must mention one cultural phenomenon that has had a most salutary unifying effect—the one issue on which all at the High Table can agree. Just as neo-Darwinians united against the (rather slender) forces of saltationism, so we disputants around the High Table in the 1980s have been united in common opposition to a movement outside the confines of biology: creationism. Indeed, the cannier creationists (and others of unmistakable creationist bent who profess simple disagreement with evolution, such as the lawyer Philip Johnson) have long accused evolutionary biologists of hiding our very real disagreements under a cloak of unanimity—so united are we against the pseudoscience of creationism.

Punctuated equilibria began showing up in creationist tracts as evidence that some scientists openly doubt evolution. The line of reasoning seems to have been: (1) Darwin founded evolutionary theory, (2) Some scientists doubt that Darwin got it entirely right, (3) Ergo, some scientists oppose evolution. Nor was this line of thinking restricted to obscure religious tracts. I am reliably informed (by the man who claimed to have pulled it off) that Steve Gould and I, thanks to our punctuated equilibria, were the scientists Ronald Reagan had in mind [sic] when he said, "Well, it is a theory, a scientific theory only, and it has in recent years been challenged in the world of science and is not yet believed in the scientific community to be as infallible as it once was believed," just after he addressed a group of fundamentalists during his first presidential election campaign.

Naturally we jumped into the fray, as did many of our opposite numbers at the High Table. Closing ranks to face a common enemy is a natural reaction. In a way, creationism was good for evolutionary biology. It made us articulate our basic precepts more clearly. And it reminded us that we have, after all is said and done, more in common as evolutionists than we have issues that drive us apart. But issues, nonetheless, there are.

Back to Biology: Punctuated Equilibria and Evolutionary Change

In its barest essentials, punctuated equilibria is a combination of empirical pattern (stasis interrupted by brief bursts of evolutionary change) coupled with preexisting biological theory (geographic speciation theory as developed by Dobzhansky and Mayr in the late 1930s and early 1940s). Punctuated equilibria has many implications, many of which provoked loud disagreements around the High Table throughout the 1980s. These involve large-scale evolutionary patterns and putative processes like species selection—the subject of the following two chapters.

Somewhat oddly, though, one of the most profound implications of punctuated equilibria has yet to excite true heated debate—at least between ultra-Darwinians and naturalists at the High Table. That is the insight punctuated equilibria yields on the critical issue of the context of evolutionary change: why evolutionary change occurs when it does. Ultra-Darwinians still blandly perpetuate the original myth that natural selection plus heritable variation plus time equals inevitable change. This is the *in principle* argument first invoked by Darwin as he contemplated the experiences of selective breeders over several generations, and imagined the outcome of a similar natural process working over the immensely longer spans of geological time. Punctuated equilibria has started the ball rolling towards a genuine understanding of why evolution happens when it does. We have finally realized that, for the most part, evolutionary change does not accumulate with the sort of reckless abandon still so strongly touted by ultra-Darwinian George Williams.

Once again we find inklings of the new insight going far back in the annals of evolutionary theory. And once again, we find it in the writings of Charles Darwin. For though he was a committed gradualist, one who freely imagined ancestral species changing through time, grading imperceptibly into descendant species, Darwin could also see the gulf yawning between closely related contemporaneous species. As if anticipating Sewall Wright, Darwin on occasion wrote of species as "permanent varieties." Like Wright, Darwin saw that the variation expressed between populations within a species is often ephemeral—subject to loss through local extinction or blending with other populations.

Not so with species. Closely related species cannot (as the overwhelming rule) blend in with their closest relatives, the way populations within a species can, and presumably often, do. Speciation—the splitting off of a portion of an ancestral species to form a new, separate descendant species—acts to conserve whatever variation and evolutionary novelties were developed in the fledgling populations, forever preventing it from loss through genetic exchange with the ancestral species. Speciation lops off a portion of the genetic variation present in the ancestral species and injects it as a separate entity into the mainstream of evolutionary history.

It is remarkable that more has not been made of this point in the 150 years of evolutionary biology that has followed Darwin's *Origin*. Speciation is critical to conserving the results of both natural selection and genetic drift. Speciation is obviously central to the fate of genetic variation, and a major shaper of patterns of evolutionary change through evolutionary time. It is as if Darwinians—neo- and ultra- most certainly included—care only for the process generating change, and not about its ultimate fate in geological time. Most of our wranglings around the High Table reflect ultra-Darwinian misunderstanding of our attempts to grapple with the fate of adaptive change through true evolutionary time. We all agree that natural selection causes evolutionary change. To an ultra-Darwinian that seems to be all that matters. We naturalists, trying to link patterns of evolutionary history up with evolutionary processes (especially selection), are trying to understand why adaptive change is developed and conserved in some cases and not in others.

What are species? Are species just collections of similar organisms, or are they something more dynamic such as collectivities of interfertile organisms? What do species do? What factors govern speciation? What causes species extinction? These are the issues that take us closer to a true link between the evolutionary process—natural selection and adaptation—and the patterns of stasis and change that we actually see in evolutionary history.

The Eternal Species Wrangle

If natural selection yields a continuum of adaptive variation, then something else must be imposing discontinuity—or so Dobzhansky, and later Mayr, reasoned over 50 years ago. That gulf between species is *reproductive*. The reason why species are "permanent varieties," able to resist melding with other species, is precisely because reproduction between members of two species does not generally occur. Sexually reproducing organisms of two separate species, no matter how closely related, don't ordinarily mate with one another across species boundaries—either because they can't or simply don't choose to even if they could.

It was logical, then, for Dobzhansky and Mayr to suggest that the source of discontinuity between species was the severing of reproductive ties between elements of an ancestral species where reproduction had formerly been unimpeded. How does such "reproductive isolation" originate? Once again, there were abundant clues from naturalists of the past. Even Darwin saw that closely related species tend to replace one another geographically. A species's closest living relative is very often found in some adjacent region, and not among those other species living cheek-by-jowl in the same general area.

That was it, then: Reproductive isolation stems from a period of enforced geographic isolation. When enough differences evolve between these disjunct divisions of a formerly single species, successful mating cannot take place if they should come into contact again at some future point. When that happens, speciation has occurred, and we now have two species where once we had one—the very gist of geographic speciation theory.

Dobzhansky and Mayr promulgated the "biological species concept" (BSC for short). It is reproductive cohesiveness that simultaneously lends integrity to a species, and keeps it separate from all others. The Dobzhansky-Mayr species concept inverted the emphasis of species definitions, which always stressed the similarities of component organisms. You still find it in school texts today: A species is a group of organisms all of whose members resemble one another more closely than they do members of other species. And, by the way, members of one species don't breed with members of another—or so goes the afterthought coda to the older, traditional, definition.

That older definition still lingers in evolutionary biology (in at least three basically distinct quarters: within the ranks of ultra-Darwinians, among some paleontologists, and among a number of cladists. What a trio!) George Williams is by no means the only ultra-Darwinian who sees no special significance for species, as distinct from any other level from populations on up through kingdoms. All are merely stages of divergence, all governed through the smooth running of natural selection-engendered adaptive change.

Some paleontologists, for their part, cling to the notion of "chronospecies"—Darwin's "insensibly graded series" which we don't see—supposedly because the fossil record is too inherently riddled with gaps. If we did see them, we would have to impose arbitrary limits between one species and its next successor as they blur into one another through evolutionary time. Phil Gingerich did, in fact, diagnose many Eocene mammal species precisely this way in the 1970s as he sought to divide his zigzagging lines of mammalian tooth evolution into separate species.

Cladists, too, tend to be unimpressed by the biological species concept. They, like ultra-Darwinians, see species as not intrinsically different in kind from any other taxon (grouping of species based on descent from a common evolutionary ancestor). It is a rare moment indeed when cladists and ultra-Darwinians see eye to eye. Such a formidable array of anti-BSC sentiment, based on an equally diverse array of assumptions and outlooks, is superficially daunting. Must we abandon the BSC, or something like it—meaning a "reproductive community" concept that features interfertility as the prime criterion of what a species is?

Absolutely not. Each of these separate cadres share a common focus: the distributions of anatomical characters in space, time, and in different organisms. An ultra-Darwinian sees in such distributions the smoothly continuous process of natural selection. So does a conservative paleontologist, hard-pressed to say anything definitive about reproduction in long-dead organisms. And the cladist is looking to map distributions of characters, to reconstruct evolutionary relationships based on the presence of shared features. The presence of a peculiar and complex internal cellular anatomy (replete with nucleus housing chromosomes), for example, unites all but bacteria into a single, unified division of life, reflecting a grand pattern of evolutionary history linking up numerous microscopic organisms with fungi, plants, and animals. To a cladist, a character may be shared by absolutely all living things (RNA, for example), or be absolutely unique to a single species—or even a single population. There's nothing unique about the species level to a cladist.

Well and good. Every field has its agenda. But not needing a concept does not invalidate it. I happen to think that cladists in fact are right. They do not need to recognize reproductive amalgams—biological species—to chart genealogical relationships. But that in itself does not mean that biological species do not exist.

I have a dimmer view of ultra-Darwinians and those conservative paleontologists who have failed to absorb the Simpsonian lesson of attention to pattern, and who continue to ignore the implications of stasis in the fossil record. To some degree ultra-Darwinians, like cladists, can afford to ignore biological species simply because their traditional—and wholly legitimate—focus of interest is on within-population genetic phenomena. But to play the *in principle*, reductive game of maintaining, as they steadfastly do, that all of evolutionary history flows smoothly from these principles, they must deny that biological species exist. This is what George Williams has done in his latest book. To an ultra-Darwinian, speciation is of no special significance beyond the simple multiplication of lineages—yielding two phyletic streams from one. In so thinking, they completely miss the point that speciation actually provides the context—the boundary conditions—for much of what really matters in terms of adaptive change in evolutionary history.

Do biological species exist? Of course they do. Here is Mayr's famous early 1940s "short" version, still memorized by many a student: "Species are groups of actually or potentially interbreeding natural populations, which are reproductively isolated from other such groups." Can we point to any such groups around us today? Of course: take our own species, *Homo sapiens*. We fit the picture, all 5.7 billion of us, however diversified we may be. All sexually reproducing organisms fit into a group within which reproduction occurs, *outside* of which reproduction does not occur.

Reproductively defined units called "species" exist in nature. They are, quite obviously, unique—as their boundaries coincide, by definition, with the limits of reproduction. No other group is defined that way. Populations within a species retain the ability to interbreed (hence their ephemeral histories—noted by Darwin, Wright, Dobzhansky, and Mayr). Larger groupings—higher taxa, such as Class Mammalia or Phylum Mollusca (clams, snails, etc.)—are composed of subunits. There may be no reproductive connections between Phylum Mollusca and Phylum Annelida (segmented worms); but neither are there reproductive connections between components of Phylum Mollusca —such as Class Bivalvia (clams) and Class Gastropoda (snails). The reproductive buck stops at the limits of a species—and it is a profound break, absolutely fraught with evolutionary implications.

Now, some biologists at the High Table have kept the notion of biological species very much alive in evolutionary discourse. They also have endeavored to discover the implications of the existence and mode of the origination of species for the understanding of the evolutionary process. Ornithologist Walter Bock, of Columbia University, comes to mind. But those of us who do grasp, in general, the importance of a reproductive concept of species in evolutionary biology do not always see eye to eye.

Bock, for example, insists that species are real, but exist only at the moment. They are real when it comes to delimiting them one against the other. But regarding species through time, Bock reverts to the standard old Darwinian assumption that, in a very real sense, species evolve themselves out of existence. The biological species concept, to Bock, has no temporal dimension—an odd position (as philosopher Marjorie Grene has remarked) for an evolutionary biologist to adopt.

But surely lineages persist through time, as reproduction goes on creating generation after generation. The reproductive chain is never broken. And, if stasis be the name of the game—if the features of organisms within species typically remain basically stable throughout the duration of a lineage—then we have every reason to see species as persisting through time. Even if there is evolutionary change, even substantial evolutionary change, lineages persist. There is a simple analogy. Each of us developed from a fertilized egg cell, grew, were born, and continued to develop and grow. We all change a great deal from conception to death, yet we are the same individual.

That's precisely how biologist Michael Ghiselin—a great admirer of Darwin, but himself only a partial ultra-Darwinian—came to see species: as "individuals." Species may undergo significant change, but they are an entity every bit as much as are individual organisms. Species have births: new species arise from old. They persist, whether more or less unchanged or not. And they eventually die, not by becoming transformed into some new species, but by simply ceasing to exist: they become extinct. We know this is so simply because the species present in the world today were not here 20 million years ago. Species in that Miocene world have become extinct, and new ones have evolved.

Species are particularized. They are discrete entities, bounded by that reproductive barricade setting absolute limits, and by discrete births and eventual deaths. And so we return to William Whewell, encountered in chapter 1, whose pronouncement gave Darwin something to shoot at. Whewell, recall, said that "species have a real existence in nature, and a transition from one to another does not exist." Darwin was determined to establish the simple truth that life has in fact evolved—that there certainly is a connection between species. In so doing, in effect he attacked the notion of species as real, and static, entities. And though the discreteness of species at times seemed an important point to Darwin, by and large Darwin neatly excised from biology the concept of species as discrete entities—feeling it necessary to do so in order to establish the validity of the notion of evolution.

It was Mayr and Dobzhansky who began the task of reemphasizing the discreteness of species. To Mayr especially, species are certainly real (if not, why have a theory—speciation—to explain their origin?). But neither Mayr nor Dobzhansky went all the way and declared species to be discrete in time as well as space. To Mayr, species were real, discrete entities at any one time. But like virtually all evolutionary biologists before him (and many since, such as Walter Bock), Mayr saw species as routinely evolving themselves out of existence once established.

Enter stasis. Species do not routinely evolve themselves out of existence, transforming like chameleons through time. Stasis is not a requirement for us to see that species have beginnings, histories, and ends, but it sure makes it easier. If all the organisms within a lineage look pretty much the same throughout their known distribution in time, it makes the job of identification much simpler. It also lends credence to the claim that we are looking at the same species throughout an interval of 10 million years or more.

Which raises another High Table issue. Ultra-Darwinians complain that it is folly to try to apply a *reproductive* concept to long-dead fossils. Paleontologist-turned-geneticist Jeffrey Levinton, from the State University of New York at Stony Brook, has been especially voluble and persistent on this very issue. Levinton claims that stasis-inclined paleontologists define species based on conveniently chosen stable characters, then turn around and shout that species are stable. How, he asks, can we be sure that our species coincide with true reproductive communities—true "biological species"?

It is certainly true that fossils seldom yield any real information about reproductive activities. None of the reproductive anatomy is preserved, for example, in any of the trilobites that I have studied throughout my paleontological career. But let's look at the problem another way: What does a paleontologist do when identifying species?

As Levinton himself knows full well, collecting fossils is very much like going to the beach. If you sort the different kinds of shells from the high-tide line, they will fall into a number of discrete piles. If you then go under the waves and retrieve living specimens, replete with all the soft anatomical parts, and with all reproductive functions intact, you will establish that your piles of shells on the beach correspond to discrete reproductive communities.

A paleontologist cannot check living samples beneath the waves. But the inference is that our samples of different shells, drawn from a single layer, conform to actual biological species as any similar sampling of beach shells would.

Then our paleontologist collects samples of shells from older and younger strata, and finds much the same sorts of shells, equally sortable into different piles. It is inference, to be sure. But species existed in the Paleozoic as surely as they exist now, and on the whole it isn't all that difficult to tell them apart.

Ernst Mayr has raised a related point: How likely is it that the earliest members of a species lineage would be able to inter-breed with members living millions of years later? Would we not assume that enough divergence had occurred so that we would have to call the early and late members of the lineage members of different species?

Mayr's point is a bit like asking how well you could relate to yourself if you, as a 50-year old, were to be taken back to meet yourself when you were three. And it is answered by Mayr's own work: the important thing is continuity. There are some species that are so far-flung and so differentiated that members of the most remote populations cannot successfully breed with one another. Yet reproductive connections exist linking them through a vast series of connecting populations. Only if those reproductive connections are severed do we say we have two species, not one. The point is not so much divergence in reproductive abilities as it is whether or not real discontinuity has intervened at some point, severing the continuum of generation-by-generation connectedness. That kind of discontinuity only arises when a species is split into two or more descendant lineages.

More High Table Flap Over Species

Though ultra-Darwinians have nearly completely turned their backs on species—their very existence, their structure, their evo-lution—by no means have all evolutionary biologists ceased to see these issues as important in evolution. Indeed, there have been several important innovations in species biology and speci-ation theory since the early days of the synthesis. One is particu-larly relevant to punctuated equilibria, and to unraveling the elusive mystery of the control and context of adaptive change in the evolutionary process. How exactly does speciation link up with adaptive change?

Part of the answer comes from the work of biologist Hugh E. H. Paterson, who has had a long and distinguished career as a field entomologist and geneticist. He returned to South Africa from Australia to run the zoology department at the University of the Witwatersrand in the late 1970s—on the proviso that the department be free of apartheid. Paterson trained a number of students now spread throughout the world, and he formulated a refinement of the BSC that has proven as irksome to some (like Ernst Mayr) as it is attractive to others.

Paterson has been especially critical of Dobzhansky's ideas on the role of speciation in evolution. Dobzhansky coined (and Mayr quickly adopted) the term "isolating mechanisms" for those traits that prevent interbreeding from occurring between two species. What are these barriers? They range from simple geographic barriers and disjunct distributions (organisms don't meet and therefore can't mate) to inherent biological properties that make breeding impossible. Thus organisms living in the same neck of the woods will meet and may attempt to mate, but their offspring will not thrive, or perhaps will be sterile. In the fullest case, organisms from two different species will meet, but won't even attempt to mate. They simply don't recognize each other as potential mates.

Paterson points out that Dobzhansky in particular saw speciation as a matter of development of these isolating mechanisms, and saw speciation as a means to keep species apart. In other words, speciation occurs so that two populations within an ancestral species can focus their adaptations more narrowly on different "niches" without worry that hybridization will dilute their ability to zero in on the precise requirements of their different habitats. Dobzhansky spoke of speciation by "reinforcement": whenever two closely related fledgling species come together, natural selection will act to intensify partially developed isolating mechanisms to prevent hybridization.

Paterson doesn't see it that way. In a subtle switch, Paterson proclaims that all those factors that tie into successful mating— meeting and recognizing a prospective mate, actual mating, successful fertilization, and production of viable offspring that will eventually mature and reproduce—constitute a single integrated system: the Specific Mate Recognition System (SMRS). This is "recognition" in the broadest sense, applying to trees and blind and rooted animals (such as corals) fully as much as lions and humans. Corals and many other groups of marine organisms shed their sperm and eggs in synchronized response to some environmental signal ingrained in all members of a particular species. This is every bit as much a matter of "recognition" as a spontaneous meeting between two passionate *Homo sapiens*.

It makes perfect sense. Reproductive systems aren't about keeping members of other species away so much as they are about finding mates and reproducing successfully. But consider the simple consequence of Paterson's SMRS concept. Speciation cannot possibly be an evolutionary process designed to partition genetic information and keep gene pools apart. Speciation must really be an accident.

In other words, natural selection simply acts to keep successful reproduction going in local populations. Speciation occurs accidentally while selection continues to act to keep that successful reproduction going. The scenario is as follows: A species becomes fragmented, so that little reproductive contact is experienced linking all its far-flung populations. The SMRS of each fragment is maintained, but they begin to drift away from one another. As selection continues to keep reproduction going within these isolated populations, divergence might become sufficiently great that successful mating can no longer occur should the two isolated groups ever come into contact again. They will no longer "recognize" one another reproductively.

Thus both sides agree that species are inherently reproductive systems. Where the BSC sees speciation as the evolution of reproductive barricades, Paterson's "recognition concept" sees those barricades as accidental fallout from the real game: continued successful reproduction within physically separated populations.

This might seem something of a distinction without a difference, and so many a biologist has said when trying to sort out what Paterson says vis-à-vis the older Dobzhansky-Mayr school.

But the distinction is there, and it is absolutely fundamental to understanding how adaptive change actually does occur in evolutionary history.

Species emerge as purely reproductive entities, and their evolution is purely a matter of (accidental) disruption of an ancestral mate recognition system into two separate systems. Contrast that with the older idea of how speciation occurs. Natural selection acts in general on separated populations. If they diverge, they may accumulate sufficient genetic differences that successful reproduction will no longer be possible should they ever encounter one another again.

What's the difference? Simple. Paterson says that only reproductive attributes—the anatomies, behaviors, and physiologies pertaining to successful reproduction—actually count in speciation. The more traditional view says speciation is a function of change in all manner of characteristics, particularly nonreproductive adaptive features—those associated with the economic side of life, such as feeding and locomotion structures. Thus the traditional answer to the query, "How is speciation linked to adaptive evolutionary change in general?," is that speciation is a direct reflection, or outcome, of general adaptive change. Let enough time go by and one species will automatically change into another. Keep two populations within a species separated long enough, and eventually you will have two species where once you had one.

But Paterson shows us that all that is necessary to get two separate species—two distinct reproductive communities—is some change in the reproductive biology of the two populations. There need be no appreciable change in the general economic adaptations of the two, just divergence in the two mate recognition systems, yielding two species where once there had been one. There is no necessary connection between speciation and (economic) adaptive change.

If Paterson is right, as evolutionary paleobiologist Elisabeth Vrba has pointed out, we should expect to see closely related species with next to no discernible differences between them. And we do: so-called sibling species, whose detection is very difficult until someone sees that reproduction doesn't occur between two sets of very similar organisms. Dobzhansky pioneered in the recognition of sibling species, finding several masquerading within his species of fruit flies.

On the other hand, we might also expect to see a great deal of adaptive differentiation within a species, yet no evidence that reproductive disjunction has taken place. Once again, we have plenty of examples of wildly varying single species where reproductive cohesion remains, but much adaptive differentiation has occurred. Savannah sparrows, distributed all over North America (and wintering into Central America), are but one example that springs to mind.

So speciation is not a simple function of adaptive change. Indeed, speciation is decoupled from adaptive change: there is no necessary relation between the two. Now how in the world can we put this together with punctuated equilibria—the profound pattern in the fossil record that seems to say just the opposite, that adaptive change is completely correlated with speciation?

There was always a paradox lurking in punctuated equilibria: Why should the occurrence of adaptive evolutionary change be so utterly dependent upon speciation? For that is what the record seems to say. You simply do not see accrual of appreciable adaptive change except when lineages divide. The fossil record tells us that speciation and adaptive change are intimately linked. How do we reconcile Paterson's recognition concept, and all it implies, with the empirical facts of the matter in the fossil record?

Speciation and Adaptive Evolutionary Change

Can speciation, in some sense, be a cause of adaptive change? The earliest critical challenge to our 1972 paper came from City University of New York paleontologist and evolutionary biologist Max Hecht, who denied that speciation, to the near-exclusion of linear, phyletic change, is the norm in evolution.

Gould and I were invited (by Hecht) to write a rebuttal to be published immediately after Hecht's comments. In it we recognized the paradox implicit in punctuated equilibria: the apparent correlation between speciation and adaptive change. We suggested a model, a causal explanation to account for the pattern of both stasis and adaptive change in association with speciation.

Once again, we relied very closely on Ernst Mayr—specifically his version of geographic speciation theory that stresses the role played by small isolated populations on the periphery of an ancestral's species range. Gould and I took up Mayr's theme and embellished it, seeking an explanation for the relatively rapid adaptive changes that seem typical of speciation patterns in general.

The idea is basically quite simple. Evolutionists of all stripes are rather fond of smallish populations, as evolution is generally seen to be more rapid in such conditions (provided the population is not so small as to preclude availability of the necessary variation). We emphasized that every species has a finite distribution in the natural world, and for good reason: species occupy terrains to which their organisms are adapted. Limits to their geographic spread are determined by the distribution of suitable habitat. It stands to reason, then, that the best habitat is located near the more central parts of a species range. The least suitable territory, as a rule, is at the margins of a species range. Indeed, ecologist George Stevens points out that many individuals on the extreme margins of a species' range are living in such suboptimal habitat, and in such reduced numbers, that frequently they are not even breeding.

Now let us suppose that some population of a species, located near the edge of that species' range in suboptimal habitat, becomes isolated from the rest of the species. We argued that natural selection, always at work to improve the fit between organism and the environment, will immediately adjust that local isolated population to fit that marginal habitat better—provided, of course, the necessary genetic variation is available for selection to work on. In a very real sense, there will be adaptive evolutionary change that redefines the habitat from suboptimal to preferred status. Unaware as we were of Paterson's insights (at that time, in the early 1970s, still in the early phase of development), we did not add that the specific mate recognition system would have to be undergoing some change for the new species to survive as a discrete entity alongside the ancestral species.

That's how we saw speciation as an actual cause of adaptive evolutionary change. The model still holds water. And although a number of biologists criticized us for placing too much emphasis on Mayr's notion of "peripheral isolates" (that is, populations isolated on the margin of an ancestral species' range), no one actually spent much time arguing against the idea. The silence across the High Table, in particular, has been deafening. Ultra-Darwinists just don't bother discussing speciation and its ramifications and connections with adaptive change all that much.

Much as I still like our original suggestion explaining how adaptive change can be seen as a result (rather than a cause) of speciation, I confess I don't think it is the whole answer to the perplexing correlation between adaptive change and speciation. There are too many different ways species can split apart—and not all offer such opportunity for rapid adaptive change to happen. Something else is afoot.

Species Selection, Species Sorting, and the Geometry of Evolutionary Change

Enter the thorniest, most provocative, and least understood element of the ultra-Darwinian–naturalist debate at the High Table: species sorting (or as it was originally and unfortunately dubbed—species selection). Species sorting is simply differential speciation or extinction of species within a larger group. Some lineages speciate at a higher rate than others, and some species are more prone to extinction than others. There are definite, repeated patterns in evolutionary history that reflect such differential speciation and extinction, patterns that had not really been

introduced until we formulated the notion of punctuated equilibria.

Species selection briefly appeared in the Great Stasis Debate
with Maynard Smith and Peter Sheldon. There, Maynard Smith
accused us of promulgating species selection as an alternative to
natural selection as a molder and shaper of adaptive evolutionary change. Our reply was that only natural selection—differential reproductive success among organisms of the same species
within a local population—could effect adaptive evolutionary
change of the basic features—the anatomies, physiologies, and
heritable behaviors—of organisms. Species selection (and species sorting) is something else, and not an alternative to natural
selection. It is a major determinant of the fate of adaptations in
evolutionary time. It is a molder and shaper, not of organismic
adaptive features, but of the patterns of both the persistence and
disappearance of those adaptations through evolutionary time.
It is time to see what species selection, and species sorting, are all
about.

Michael Ghiselin and philosopher David Hull have argued
strenuously that species are particularized entities—"individuals" in the same sense that an organism, a state, or a corporation
is a large-scale entity. Species have names; they have beginnings,
histories, and ends. They are discrete historical entities.

It is not surprising, then, that some species last longer than
others. Extinction occurs at faster rates in some lineages than
others; so, too, with speciation. Speciation rates are faster in
some lineages than in others. There are, it follows, factors that
bias the births and deaths of species—that bias speciation and
extinction rates of species. In principle, such biases would be
expected to leave their mark on the exact shape of evolutionary
history—not on which adaptations appear, but on what the fate
of those adaptations might be in evolutionary time. As it turns
out, the explanation of the paradox hinges on this very issue.
The paradox, once again, is that adaptive change seems so completely bound up in speciation events; even though we know
there should be no necessary correlation between reproductive
and adaptive change in evolution. Its resolution lies in the fac-

tors that bias rates of speciation and extinction in evolutionary history.

We know we cannot be seeing all the speciation events that have ever occurred, even in lineages that have left a relatively good, dense fossil record. That's because there are sibling species, whose organisms are reproductively discrete, but otherwise just about identical. But sibling species, in an important sense, are irrelevant when it comes to understanding what's going on with adaptive change in evolution. By definition, sibling species show next to no adaptive change. We want to know why and when adaptive change does occur.

Naturalists espousing enthusiasm for punctuated equilibria have taken quite a lot of heat for suggesting that speciation causes adaptive change. Johns Hopkins paleobiologist Steven M. Stanley (who coined the term "species selection"—at least in the sense used here) has born the brunt of the criticism for espousing a hard-core form of speciation that he saw as necessarily entailing significant amounts of anatomical change. But all of us have been accused of claiming that speciation invariably produces great amounts of adaptive change as part of the charge leveled against us that we are barely disguised Goldschmidtian ("Gouldschmidtian") saltationists.

In fact, only when rhetorical usage finds it convenient do ultra-Darwinians admit that we naturalists see an entire spectrum of change associated with speciation, from virtually none to quite a lot. Maynard Smith, for example, acknowledges that in our original examples of punctuated equilibria, we documented only minor amounts of anatomical change in our speciational events. His point was to find fault with our charge that Sheldon's trilobites, evolving little over their three-million-year interval, have little to tell us about how adaptive change actually accumulates in evolutionary history. Yet our examples showed small but measurable, and persistent, change over a brief spurt of time. Sheldon's lineages took vastly longer amounts of time to waffle about. There *is* a difference.

Nonetheless, the standard ultra-Darwinian charge is that we naturalists imbued with punctuated equilibria invariably see large leaps occurring rapidly between our species. What we do claim to see is that most persistent adaptive change is correlated with speciation events. What's going on?

We are seeing, I firmly believe, only successful speciation events. Consider the common fate of fledgling species. I was, as a student in the 1960s, immensely influenced by the work of University of California (Berkeley) botanist Harlan Lewis. Lewis worked on the genetics of the plant genus *Clarkia*. In some populations he discovered that a fairly severe mutation occurred at sufficiently high frequencies, that each year some small populations would be established that were reproductively disjunct— isolated—from the parental species populations. Instant speciation (although I remain content with the "five to fifty thousand year" estimate, which is by far the more usual rate—a rate that only *looks* fast when viewed against the enormity of true evolutionary time).

What was the fate of these little disjunct *Clarkia* populations? Extinction habitually claimed them all. They may well have been reproductively isolated from the parental species, but they remained ecologically identical. Unable to gain a distinct foothold, they were literally swamped out by the vastly greater numbers of the parental species. The moral of the story: Fledgling species have a much greater chance of survival if they are ecologically differentiated from the parental species.

It is no great trick, following Hugh Paterson's arguments, to see new species appear as the SMRS diverges in isolated populations. But fledgling species have, if anything, a greater likelihood of becoming extinct than well-differentiated populations within a species—unless some degree of ecological differentiation accompanies the change in reproductive adaptations. We are seeing the survivors in the fossil record; they are the new species that were ecologically distinct from their ancestral species—at least sufficiently to allow them to establish an ecological foothold and to persist. And their ecological distinctiveness is mirrored in their anatomies, which bear the traces of evolutionary, adaptive change.

Speciation, after all, does not necessarily beget evolutionary adaptive change. But *successful* speciation—persistence past infancy, survival long enough to show up in the fossil record—does. That's why adaptive change is so tightly correlated with speciation. Without speciation, little change is possible: stasis is the rule. With speciation, there may be the opportunity for rapid evolution. If not, if little adaptive change occurs at speciation, we get high rates of fledgling species extinction—and no insight about why adaptive change occurs when it does. But if some change does occur, we get a much greater chance for species survival and the production of still more new species down the phylogenetic road. Thus is adaptive change injected into the evolutionary stream.

That's it, in outline form. There's a built-in bias involving survival rates of fledgling species. Fledgling species are being sorted in a systematic fashion. Species sorting does not determine what new adaptations appear in evolution; rather, species sorting depends on whether or not such adaptive change occurs—through normal Darwinian processes. But the effect of that sorting is monumental in shaping the overall course of evolutionary history. It is that sorting that lies at the heart of the paradox: why so much of adaptive evolutionary change is tied in with speciation.

That, then, is the central message of punctuated equilibria. We can understand patterns of stasis and change using familiar biological concepts, certainly including within-population natural selection, but also using modern concepts of species and speciation. And we see, as a consequence, that species are real, historically bounded entities—entities that can be sorted, or even selected.

The plot thickens. For there are many other implications of punctuated equilibria, all hinging on the importance of species as a level of biological organization that cannot be ignored or dismissed in a reductive wave of the hand. There is much more to the species selection/species sorting story. There are large-scale evolutionary patterns, so-called macro-evolutionary patterns, that have surfaced in the punctuated equilibria debate, prompting the loudest howls of outrage from the ultra-Darwinian side of the High Table.

5

High Stakes at the High Table

Macroevolution and Species Sorting

If punctuated equilibria and the issue of stasis brought paleontologists back to the evolutionary High Table, easily the most provocative conversational gambits of the ensuing discussion were two essays by Steve Gould in 1980 and 1982. The earlier of the two modestly asked (in its title), "Is a new and general theory of evolution emerging?" To Steve, the answer was clearly yes. The synthesis, he declared, was effectively dead, replaced by a newer theoretical structure we naturalists had been formulating. We had come to see that species and higher taxa exist as discrete historical entities, and have their own differential evolutionary fates that cannot be reduced simplistically to within-population processes. We were beginning to see that ecosystems and social systems are also large-scale entities not satisfactorily addressed in traditional evolutionary theory. We had begun to take the hierarchical structure of biological nature seriously.

Ultra-Darwinians, quite naturally, reacted with outraged incredulity at Steve's pronouncement. Announcing the death of the synthesis was waving a rhetorical red flag so provocative that ensuing discussion was more visceral than cerebral. Ultra-Darwinians quickly mounted a countercharge: that we naturalists (Gould in particular) were trying to throw out natural selection, only to replace it with an unfounded concept of "species selection." English geneticist John Turner, when calling our naturalist position "evolution by jerks," claimed that there is "no good evidence" for the central ideas of punctuated equilibria: "I am tempted to say no evidence at all," said Turner in 1984.

At stake was the entire in-principle stance of Darwinian evolutionary theory—where all of evolutionary history could be understood as flowing smoothly and directly from the generation-by-generation process of adaptation through natural selection. With the publication of our 1972 paper on punctuated equilibria, paleontologists once again began to pick up the Simpsonian theme that large-scale patterns in evolutionary history cannot be explained as a simple, in-principle, fallout of business-as-usual natural selection within local populations. Ultra-Darwinians were wrong when they asserted that we were throwing out the body of Darwinian insight on adaptive change through natural selection with the bathwater of reductionism. But we definitely were saying that the large-scale patterns of the history of life, such as evolutionary trends and the relation between extinction and evolution, demanded additional explanation. This additional theoretical treatment must go beyond the simple handwaving extrapolationism that to this day forms the very heart of ultra-Darwinian treatments of "macroevolutionary" large-scale evolutionary events and patterns.

The very term *macroevolution* is enough to make an ultra-Darwinian snarl. Macroevolution is counterpoised with *microevolution*—generation by generation selection-mediated change in gene frequencies within populations. The debate is over the question, Are conventional Darwinian microevolutionary processes sufficient to explain the entire history of life? To ultra-Darwinians, the very term macroevolution suggests that the answer is automatically no. To them, macroevolution implies the action of processes—even genetic processes—that are as yet unknown but must be imagined to yield a satisfactory explanation of the history of life.

But macroevolution need not carry such heavy conceptual baggage. In its most basic usage, it simply means evolution on a large-scale. In particular, to some biologists, it suggests the origin of major groups—such as the origin and radiation of mammals, or the derivation of whales and bats from terrestrial mammalian ancestors. Such sorts of events may or may not demand additional theory for their explanation. Traditional Darwinian explanation, of course, insists not. But George Simpson, as I have already noted, felt that the abrupt Eocene appearances of both bats and whales indicate a special period of rapid evolutionary change that demands detailed theoretical explanation.

Indeed, Simpson, noting that microevolution refers to within-species processes, while macroevolution refers to relatively lower-ranked higher taxa (meaning genera and perhaps families), decided that still a third term would be useful: *megaevolution*. This term would embrace the very sorts of examples of large-scale evolution he had in mind with his concept of quantum evolution. I think it is safe to say that all evolutionary biologists agree with Simpson's later decision to abolish the use of megaevolution (no one had picked it up, anyway). Macroevolution is confusing enough.

Though Simpson's quantum evolution addressed the origin of large-scale groups of organisms—so-called higher taxa—he was, in reality, discussing large-scale adaptive change. Whales and bats are natural groups, consisting of hundreds and thousands of species over the past 55 million years. But it was the aquatic and aerial adaptations of those animals, adaptations that define the mammalian Order Cetacea (whales) and Order Chiroptera (bats), that were really at issue from the standpoint of evolutionary process. Simpson's quantum evolution was a straightforward attempt to specify the special conditions that must obtain in order for microevolutionary processes to produce macroevolutionary effects.

That is really all we naturalists have been trying to achieve: to formulate the circumstances and context in which microevolutionary change works in evolutionary real time. We have elaborated only one significant piece of additional theory: species sorting. As we saw in chapter 4, drawing even on a part of Darwin, and certainly on the pioneering efforts of Dobzhansky and Mayr, we see species as real historical entities. They have beginnings (in speciation), histories, and ends (extinction). We see that adaptive change is intimately bound up with the origin of new species. Once a new species appears, the adaptations of its component organisms are injected into the evolutionary mainstream. Species are packages of genetic information—information pertaining to the functional properties of its organisms. It stands to reason, then, that if some species appear faster than others in a lineage, or if some become extinct more readily than others, there will be a juggling, a sorting, of species, and thus of the genetic information they contain. Anything that biases the births and deaths of species necessarily biases the fate of adaptations and the genetic information on which they are based.

Thus species sorting is not a new, or rival, theory of evolutionary adaptive stasis and change. It is additional theory that helps us understand why adaptive change seems caught up with the origin of new reproductive communities: new species. It is an enrichment of evolutionary theory, an expansion of traditional discourse. It could not be otherwise. Natural selection is the effect of differential reproductive success of organisms within populations. Natural selection biases the fate of genetically based variable traits of organisms within populations. The domain of natural selection is within populations, within species. Species sorting, also by definition, does not work on heritable variation within species. Its domain is between species; species are sorted via differential rates of their production and extinction. Species sorting affects the fates of adaptations once they are evolved. The two ideas are complementary—not rivals, as Maynard Smith, Richard Dawkins, and other ultra-Darwinians insist.

A Punctuated Paradox: Evolutionary Trends and the Origin of Species Selection/Sorting

Paleontologists have always been fascinated by long-term trends in evolution. From the early decades following publication of Darwin's *On the Origin of Species* to present times, paleontologists have continually cited examples of evolutionary change continuing in a single apparent direction over truly prodigious periods of geological time. Such evolutionary trends have proven to be a tortuous, tricky, and increasingly tendentious subject.

To early paleontologists heeding Darwin's call to produce "insensibly graded series" of fossils from the geological strata, it was a matter of course that series of fossils trending in one particular evolutionary direction would soon turn up. If it proved difficult to produce layer-by-layer finely drawn examples of directional change—such as Darwin and nearly everyone else expected—eventually long-term directional change would shine through the gappiness of the fossil record. And paleontologists had no trouble producing many examples.

The most famous of all evolutionary trends produced in the initial burst of evolutionary-inspired research came from Russian paleontologist Vladimir Kovalevsky, who pronounced the European Eocene paleotheres as early ancestors of modern horses. The pioneering explorations of Tertiary mammals from the American West soon made it clear, however, that the vast bulk of horse evolution had taken place in the New World. When another early student of fossil horses, Thomas Henry Huxley (Darwin's "bulldog"), came to the United States in 1876 on a proselytizing speaking tour, he made a beeline to Othniel Charles Marsh's laboratories at Yale University. Marsh, spurred on by an increasingly bitter rivalry with Edward Drinker Cope of Philadelphia, had amassed a great array of Mesozoic and Cenozoic fossils, among them a number of ancient members of the Equidae, the horse family.

Huxley just had to see those fossils, with their ultimate vindication of the Darwinian evolutionary picture. For even though he had admonished Darwin not to press the point that "Nature does not make leaps" too strongly, horses seemed to display just the kind of overall change through time that Darwinian evolutionary theory predicted.

Marsh's horse fossils did not disappoint. True, no longer could Huxley cling to the simple, linear picture of horse evolution restricted to the Old World. Marsh convinced him that the true ancestry, and most of the subsequent evolutionary history, of horses actually took place in the Americas. But the increasingly rich picture of horse evolution is a feast for any evolutionary biologist's eyes. As generations of school children have learned from exquisite exhibits at the American Museum of Natural History, reproduced in textbooks throughout the world, horses changed quite a bit from their earliest days around 60 million years ago. They got bigger. The toes on both front and hind feet were reduced in number from four in front and three in back in *Hyracotherium*, the earliest ("dawn") horse, to the single digit of modern zebras, asses, and horses. Teeth switched over from lobed molars suitable for browsing, to many-ridged teeth suitable for grazing the highly siliceous grasses that appeared in the Miocene. (Opal is a form of silicon dioxide, as is glass. It is the little particles of "glass" on the edge of a blade of grass that can cut your finger open.) And those grazing teeth got taller, longer, and more complex as time went on.

The museum's old display (recently revamped) included four near-perfect skeletons arranged in order of geological occurrence, from Eocene to the present. The changes are strikingly apparent. At the height of the creationist controversy in the early 1980s, I went on TV, using the horses as a backdrop to show the truth of the general evolutionary assertion that organisms indeed do change through time. Creationists, in turn, used the horses to depict how nefarious evolutionists can be. They charged, completely falsely, that we evolutionists had lined up the fossil horses to demonstrate evolutionary change, without regard to their proper placement in geological time. Nothing could be further from the truth. Those horses were lined up in the exact order of geological sequence in which they were discovered.

But trends are indeed a tricky subject. George Gaylord Simpson spent a considerable segment of his career on horse evolution. His overall conclusion: Horse evolution was by no means the simple, linear and straightforward affair it was made out to be. Yes, those fossils on display reflect the true position of four species in geological time. But horse evolution did not proceed in one single series, from step A to step B and so forth, culminating in modern, single-toed large horses. Horse evolution, to Simpson, seemed much more bushy, with lots of species alive at any one time—species that differed quite a bit from one another, and which had variable numbers of toes, size of teeth, and so forth.

In other words, it is easy, and all too tempting, to survey the fossil history of a group and select examples that seem best to exemplify linear change through time. The net evolutionary change, after all, is real. But picking out just those species that exemplify intermediate stages along a trend, while ignoring all other species that don't seem to fit in as well, is something else again. The picture is distorted. The actual evolutionary pattern isn't fully represented.

It is easy to see how a penchant for simplistic linear renderings of real examples of evolutionary change originated. It is the very essence of Darwinian extrapolationism to imagine that long-term trends are simply shorter term, within-species examples of linear, selection mediated trends. Put more positively, there is a seamless connection between short-term microevolutionary change projected up through geological time to embrace tens of millions of years of (purportedly) directional evolutionary change.

The evolutionary change in all these examples is real, and it is directional: net increase in size, net decrease in number of toes in horse evolution. Perhaps an even more gripping example comes from human evolution. Four million years ago, early hominid brains averaged something like 450 cubic centimeters (milliliters). Our own brains average out at some 1,400 cubic centimeters. Samples drawn from intermediate strata throughout the past four million years are themselves pretty much intermediate, if we ignore some of the side branches whose brain sizes fall off the nice neat line of gradual increase through time.

But if the trends are not exactly linear, they are nonetheless real. There is net direction in many evolutionary lineages. Trends are indeed real, however simplified to fit a single linear picture they may have been rendered. And that was a real challenge to us as we developed our ideas on punctuated equilibria.

Stasis really throws a wrench in the neo-Darwinian works. It is one thing to proclaim that stasis, rather than smooth linear directional change, is the norm within species. But if we acknowledge that long-term trends, albeit bushier than generally conceded, do constitute a major category of pattern in evolutionary history, how to explain them? Stasis deprives us of the time-honored explanation: long-term trends are simply reflections of shorter term, within-species microevolutionary trends. But if we see no evidence for those trends within species—if stasis is the norm—how is it that there often seems to be long-term, between species ("macroevolutionary") net directional change?

That was the paradox we faced from the outset as we developed the idea of punctuated equilibria. Stasis is the rule within species. Horse species change but little over their average durations of a few million years. And the fossil record of hominid species, spotty as it is, shows great stability in brain size. The mid-Pleistocene species *Homo erectus*, according to paleoanthropologist Philip Rightmire, had a brain size that remained stable at about 1,000 cubic centimeters for more than 1.3 million years. Yet it sits right in the middle of an overall trend of brain size increase that stretched from 450 cubic centimeters four million years ago to our own 1,400 cubic centimeters.

We had a problem on our hands. Just as Simpson felt that evolutionary patterns in the fossil record demanded additional theoretical treatment, Gould and I proposed (toward the end of our original punctuated equilibria paper in 1972) that some sort of bias in the survival of different species within a lineage would secondarily act to impose a sense of linearity, or directionality, to a lineage. We proposed a loose analogy with natural selection, postulating that adaptive change occurring at speciation need not go in any particular direction. Brain size, for example, might as likely decrease as increase; or horse toes might increase as well as decrease. Whatever change occurred, it would be under the general aegis of natural selection (and perhaps genetic drift). Nothing particularly shocking here.

What renders the trend, then, is differential species survival. The species that happen to develop adaptive change in one particular direction are the ones that tend to survive, to produce further species, thus handing down to descendants those very features that conveyed success on themselves. We proposed a sort of pruning process, in which only some of the many species extant at any one time would persist and give rise to descendant species.

Note that nothing in this proposition of differential species survival pertains to the actual *origin* of adaptive change per se. Rather, we were addressing only the relative probability of survival of different packages of genetic information—species— each of which are composed of organisms with slightly different adaptations. However, by far the weakest part of the proposition was the analogy that we drew between direction of adaptive change at speciation and the process of mutation. Mutations have definitive, deterministic causes. They are essentially biochemical mistakes, most often arising from the molecular process of duplication or translation. Whatever their cause, though, all evolutionary biologists agree that mutations are random with respect to the needs of the organisms in which they occur. A mutation might be harmful, beneficial, or quite neutral to the organism that carries it. Mutations just happen, and certainly do not arise because they might be beneficial to the organism in which they happen. Natural selection works to eliminate harmful mutations after the fact, keeping beneficial ones and ignoring neutral ones.

In contrast to our speculation that adaptive change in species might likewise be random with respect to long-term evolutionary trends within a lineage, there is no evidence that human brain size ever became smaller when a new species arose—or that horse toe numbers increased when some new species appeared. One could imagine that such events might have happened, but that such endowed fledgling species simply died out. But there is really no evidence forthcoming to support such a fantasy. Rather, it is far more likely that natural selection never did produce a hominid species with a smaller brain than its direct ancestral species. The better model, I believe, is to posit that at speciation there are two possibilities: no change at all in the trait that shows the trend, or natural selection modifies adaptations in the direction of the overall trend. Speciation, with natural selection–mediated adaptive change, just can't be exactly analogous to mutations.

Other evolutionary biologists have, from time to time, proposed a form of directional speciation. Whenever speciation occurs, it always produces change in the overall direction of a trend. This is a far stronger form of our proposal, and one that I have difficulty accepting because I (along with ultra-Darwinians) have great difficulty seeing why this should be so. But I have no doubt that some form of differential species survival takes place within lineages, and that trends—net changes in some important trait within a lineage of many species—come about through this form of species sorting.

Mass Confusion at the High Table: Species Selection, Species Sorting, and Group Selection

One of the earliest and strongest backers of what he called the "punctuational model" was paleontologist Steven M. Stanley. In the mid-1970s, Stanley was in his early 30s, but had already coauthored an influential textbook on paleobiology with David M. Raup, a leader in the new movement to inject more and better biology into paleontological theory and practice.

In 1975, Stanley published a paper entitled, "A theory of evolution above the species level," in the prestigious *Proceedings of the National Academy of Sciences*. Here he coined the term "species selection" to embrace the kind of biases in species persistence that Gould and I had invoked to reinstate directionality in evolutionary history. Stanley went on to publish an important book, *Macroevolution: Pattern and Process* (1979), and has continued throughout the 1980s and early 1990s to develop the naturalistic perspective. His documentation of the pervasiveness of stasis in the fossil record, as well as his analysis of mass extinctions, have been especially noteworthy.

There have been two distinct debates over the nature and efficacy of species selection throughout the 1980s. On the one hand, we naturalists have been at odds with one another about the precise definition, scope, and power of species selection. On the other hand, we are united in our insistence that, whatever the precise nature of the process, species are indeed bounded entities with differential births and deaths, and that such a process has profoundly shaped the course of evolutionary history.

Ultra-Darwinians disagree. Part of their resistance stems from a confusion over a raging debate going on within their own circle. This is their own internal argument over the nature and reality of what they call "group selection"—the proposition that adaptations can arise as a benefit to entire populations, rather than simply to the benefit of individual organisms.

George Williams, recall, got the ultra-Darwinian ball rolling with the publication of his *Adaptation and Natural Selection* (1965). It was a thoroughgoing defense of the efficacy of natural selection as the molder and shaper of evolutionary change. It was written, at least in part, to refute the heresy of his day: Australian biologist V. C. Wynne-Edwards's notion of group selection. Wynne-Edwards served as the same sort of lightning rod to Darwinians as Richard Goldschmidt had a generation earlier. Wynne-Edwards denied that natural factors—energy resources, disease, climate, and so forth—were the factors acting to limit the size of populations. He proposed instead that, particularly for social animals, such regulation of population size stemmed from the organizational aspects of populations themselves. He felt sure that the population size and structure of each different species reflected the optimum for that species: it was for the "good of the species" that population size was regulated.

To any good Darwinian, saying that something is "for the good of the species," rather than for the good of the individual, is a fundamental heresy. Yet it has been population geneticists, well ensconced within the neo-Darwinian fold, who have themselves performed experiments and theoretical mathematical analyses, examining the previously unthinkable proposition that such a thing as group selection might exist.

The debate over group selection in population genetics has little or nothing to do with our own internal debate over species selection. But, of course, species are groups, and therein, I think, lies part of the confusion that has plagued discussion around the High Table. For insist as we may that species selection has nothing to do with the origin and further change of heritable traits, ultra-Darwinians don't believe us. They don't believe us simply because their own debate on group selection is about adaptive change. They see group selection as an antithetical, direct competitor with natural selection as an explanation for adaptive change. We see species selection as additional theory that explains large-scale evolutionary patterns, such as trends, but does not compete with natural selection as a mechanism that generates adaptations.

Group selection involves heritable organismic traits that benefit a group—a deme, say, or for that matter an entire species. Most debate over group selection in modern Darwinian circles deals with situations where a heritable characteristic may be favorable to a group as a whole but disadvantageous to organisms taken singly. In such an instance, group selection is in conflict with natural (individual) selection. The question to an ultra-Darwinian is, "How important a consideration is group selection in understanding the evolution of organismic traits?" And, though group selectionists have managed to establish a toehold of respectability in ultra-Darwinian circles, in general the Williams-Dawkins implacable resistance to anything but individual selection (gene selection, to Dawkins) has resisted any truly serious intrusion of group selection into the core neo-Darwinian paradigm.

In contrast, no discussion of species selection within the naturalist camp has ever invoked "relative fitness values," or any other concept that would suggest that we are guilty—like Wynne-Edwards—of promoting anti–natural selection heresy.

Ultra-Darwinians seem to have made no clear distinction between their own debate and our own internal wranglings.

And wranglings there have been. Most clearheaded in all the confusing interchanges, in my opinion, have been the contributions of paleobiologist Elisabeth S. Vrba, formerly of the Transvaal Museum in Pretoria, and for the last decade at Yale University. Vrba saw no particular difficulty in making an analogy between natural selection, on the one hand, and something that might be called "species selection" on the other—so long as the analogy is a proper one. I am convinced, on the basis of Vrba's analysis, that we naturalists have been saying "species selection" when we should really have been calling the phenomenon "species sorting." Species sorting is extremely common, and underlies a great deal of evolutionary patterns, as I shall continue to make clear in this narrative. On the other hand, true species selection, in its properly more restricted sense, I now believe to be relatively rare. Here is why—once again, following Vrba's analysis.

First of all, Vrba pointed out, only in some of its primitive incarnations was natural selection ever seen strictly as an eliminative process. Natural selection involves differential births as well as deaths, and is treated purely as differential reproductive success in modern Darwinian circles. In contrast, she said, Eldredge, Gould, and Stanley addressed only differential species deaths. Surely, she argued, any true concept of species selection should also address differential species births as well as their deaths.

Then, too, natural selection is a within-population process that shapes adaptive stasis and change from available heritable variation. To be a true analogue to natural selection, species selection would work on the properties of species, and not on the properties of organisms. And, though our early versions of species selection did not address the origins or modifications of adaptations, they were addressed to the fates of organismic traits and not the properties of species per se.

Touché. Do species have properties of their own? Yes, they do, but they are not items like brain size or number of toes on the front foot. Species have distributions in space and time. Paleon- tologist David Jablonski has suggested that geographic ranges are indeed heritable, and thus a possible focus of species selec- tion. But somehow such ranges don't seem truly comparable to the intrinsic, genetically based traits of organisms.

It was Maynard Smith who first suggested to me (in a pub in Manchester, England) a more legitimately intrinsic property of a species. I have come to adopt this view, especially at the prod- ding of several of my students. Maynard Smith pointed out that sexual reproduction, involving as it does participation of one organism from each of two sexes, in itself constitutes a species- level property. Hugh Paterson's notion of the Specific Mate Rec- ognition System, the fertilization system unique to each species, then serves as a true species-level property. This has some inter- esting theoretical implications that are pursued in the following chapter.

But I firmly believe there are no other legitimate species- level properties. Vrba's arguments convinced me that "species selection" is a misnomer. She suggested a more neutral term— species sorting—for instances in which it is unclear whether dif- ferential species births and deaths reflect true species selection or some other cause. Maybe, she suggested, species could be sorted simply as a result of the properties of their component organ- isms. I discuss her original example, involving two related groups of African antelope, in the next section.

That's why I used the term "species sorting" in the previous chapter. Probability of survival of fledgling species, I claimed, is a matter of how much adaptive differentiation has taken place: in other words, how different the organisms of the fledgling spe- cies are from those of the ancestral species. The term "species sorting" has the added benefit of dropping the confusing term "selection," possibly damping the false ultra-Darwinian charge that we are hatching alternate theories to natural selection. But perhaps not. "Species sorting" has been bandied around the High Table for nearly a decade now, and so far, ultra-Darwinians still seem to think of it as a wolf in sheep's clothing—a simple cover-up for the dreaded species selection.

Vrba's "Effect Hypothesis"

Elisabeth Vrba works on African antelopes—including modern species, but especially extinct species that have left interesting evolutionary patterns in the fossil record. She has made a number of crucial contributions to evolutionary discourse at the High Table, all of which stem from her documentation of two basic patterns. The first involves evolutionary trends, the familiar pattern which led to her novel suggestion of the "effect hypothesis." The second bears on extinction as an ecological phenomenon, and its relation to bursts of evolutionary change: Vrba's "turnover pulse hypothesis," another major contribution that sheds critical light on the environmental context of evolutionary change. But first things first.

As a systematist, Vrba has worked intensively to clarify the evolutionary relationships among antelope species. One of her early conclusions suggested a novel alignment between impalas and a group of unusual species centering around the two wildebeest (gnu) species, the blue and black wildebeests. Gnus are by all odds the most highly specialized of antelopes—with long-muzzled faces that give an appearance far removed from typical antelopes. They and their closest relatives (hartebeests and topis) are ecological specialists, closely dependent on narrow habitat requirements. Species of the wildebeest group (Aepycerotini) typically rely on a narrow range of plant species for forage, and are likewise strongly dependent on reliable water supplies. The great wildebeest migrations of the African savannahs reflect their constant need to fulfill these highly particular requirements as rainy seasons alternate with times of drought.

Impalas are another story altogether. There are some six or seven species in the wildebeest group, but only one impala species. Impalas range over vast segments of the African landscape, occupying a wide range of habitats. Not dependent on any single plant species, nor restricted by constant reliance on ready availability of water, impalas are ecological generalists. Physically, impalas are very generalized antelopes, retaining many primitive features. We begin to see a correlation emerging. Ecologically specialized species possess many anatomical features—adaptations—that are themselves evolutionary specializations. Ecologically generalized species tend to retain many of the primitive features that were present in remote evolutionary ancestors.

Specialist versus generalist. It's an interesting contrast; one that links ecology with a particular form of evolutionary pattern that crops up over and over again as we survey the fossil record. As long ago as the 1880s, Cornell University paleontologist Henry Shaler Williams (looking primarily at Devonian brachiopods) noticed that there are two sorts of species in the fossil record. One type seems to show up all over the place in a wide variety of environments, suggesting relatively great environmental flexibility. These species, like Vrba's impalas, seldom feature distinctive novel adaptations. They vary a bit in place to place (as do Vrba's impalas—mostly in overall size), but there is seldom any major adaptive differentiation to be seen in such species.

There is one absolutely critical aspect to such ecologically and adaptively generalized species, an aspect that only a paleontologist could see. Such species typically last a very long time. Ecologically and adaptively generalized species tend to be much more highly resistant to extinction than specialized species. Williams had many such specialized species in his Devonian brachiopods. Invariably restricted to a few different environments, and often sporting distinctive anatomical specializations, such species as the overwhelming rule do not last nearly as long as their more ecologically generalized brethren.

The pattern makes great intuitive sense. Extinction, as we are seeing with increasing clarity and alarm in the modern world, is a function of habitat degradation and loss. It stands to reason that the more ecologically generalized a species is, the greater its chances will be to continue to recognize habitat in a changing world. This is the phenomenon of habitat tracking so closely connected not only to simple persistence, but to the very phenomenon of stasis.

In contrast, specialized species, focused as they are on far narrower spectra of habitat needs, are sitting ducks. Unable to switch to alternate food sources, less able to withstand the physical rigors of alternative environments, specialized species put all their ecological eggs in a single basket and risk extinction if that habitat should disappear.

Why should some species consist of specialists and others of generalists? Put another way, why do the organisms of some species seem to perceive their environments in a very narrow, fine-grained manner, while the generalists have a very coarse perception of their habitat?

Vrba and ecologist Michael Greenacre looked at census figures of antelope individuals in various African game parks and saw that there are as many pounds of impala flesh out in the wild as there are pounds of all seven species of Aepycerotini combined. That's very interesting. It suggests that one adaptive strategy at any given moment is as good as the other. It shows that nature accommodates far fewer generalist than specialist species in any one area, but hosts as many generalist as specialized organisms within those species.

The advantage of specialization is efficiency: focusing on a single, presently abundant and reliable resource makes perfect adaptive sense. George Williams was absolutely right when he insisted (in *Adaptation and Natural Selection*) that natural selection can have no "eyes" for the future. There is no way natural selection will not work to fashion narrow, specialized adaptation because the entire species is put at higher risk of eventual extinction. What is important in evolutionary adaptation is the moment-by-moment "what works in the here and now," and not some unanticipated collapse of the habitat on which the specialist utterly depends.

Henry Shaler Williams's pattern—the correlation of ecological and adaptive generality with extinction resistance—has been found over and over to hold true. Vrba saw it in her antelopes, and I have seen it in trilobites. In 1984, Steven M. Stanley and I edited a volume of papers addressing the general phenomenon of "living fossils"—the persistence of ancient anatomies for often prodigious periods of geologic time. The horseshoe crab, *Limulus polyphemus*, looks an awful lot like horseshoe crabs that were scuttling around the sea floor in the late Paleozoic over 250 million years ago.

Looking for causal explanations of this hoary phenomenon, many of our authors demonstrated the same sort of pattern displayed by Vrba's antelopes. Often, the closest relatives of the ancient, persistent types is a lineage consisting of many more species, all showing evolutionary specializations of one form or another. All of these specialists, sooner rather than later, succumb to extinction while the primitive types keep plodding on. It is very much a tortoise and hare situation. The ecological specialists spurt ahead, developing evolutionary adaptive specializations, but, more often than not, the specialists fall by the wayside.

Stasis within species is the norm whether the organisms are highly specialized or extremely generalized in their adaptations. With the Henry Shaler Williams–Elisabeth Vrba pattern, we move up a step. We are now comparing relative amounts of evolutionary change between lineages. Lineages consisting of species who remain generalized have fewer species, tend to resist extinction, and accumulate little or no major adaptive change. Vrba counts just two species of impala over the past six million years, while there have been a minimum of thirty species in the wildebeest group over that same span of time. These patterns, then, are macroevolutionary, and involve the differential fates of species within lineages.

But there is more to the pattern of specialized versus generalized lineages than simple differential rates of extinction. There is also the striking matter of differential rates of speciation, and the correlation of high rates of speciation in the specialized lineages with the accumulation of vastly greater amounts of adaptive evolutionary change. Recall that Vrba's wildebeests are among the most specialized (even downright anatomically bizarre) of all antelopes—fossil or modern.

Indeed, looking at the sequence of evolutionary change within the wildebeest group, Vrba saw a trend towards increasing specialization and extreme anatomical configuration as geological time rolled by. Once again, the trend is not a reflection of gradual, linear directional change within species. Rather, it is punctuated: change occurs in association with speciation. Once a new species appears, the old familiar pattern of stasis takes over. The trend, once again, is an accumulation of adaptive change between species within a lineage as geological time rolls by.

But Vrba found an additional ingredient in her pattern. Rather than a pruning of species guiding the accumulation of adaptive change in a particular direction, she saw that the rate of speciation itself was higher among the wildebeests than the impalas. Ecological specialization, it seems, confers higher rates of speciation, and thus of development and accumulation of adaptive change—along with the more familiar high rates of species extinction.

If it is relatively easy to grasp why specialized species are more extinction-prone, it is not immediately obvious why speciation rates are also high in lineages of specialists. Speciation, after all, occurs when reproductive adaptations diverge in isolation. And there is little reason to suppose that generalized species are any less prone to disruption of the "specific mate recognition system" than specialized species. In other words, it is not immediately obvious why ecologically specialized species should be more prone to become fragmented into reproductively isolated new species than their more generalized relatives, or so I tend to think. Vrba herself suggested precisely that. Ecologically specialized, narrow adaptations go hand in hand with a tendency for adaptive modification of the mate recognition system—speciation.

This is an important issue, and a source of some disagreement among those of us in the naturalist camp. What underlies the correlation between ecological specialization and relative higher rates of speciation? I see the problem, once again, primarily as a matter of survival of fledgling species. My explanation for the correlation between adaptive change and speciation (developed in the preceding chapter) hinges on the supposition that the greater the ecological difference between parental and fledgling species, the greater the chances of that fledgling to survive and become established as a successful, independent species.

With the pattern exemplified by Vrba's African antelopes, where it is the more ecologically specialized species that show higher rates of origination, we see the very same principle in extended form. The argument is as follows: with broadly adapted species, a newly isolated fledgling species must quickly develop a relatively great amount of adaptive change—taking it outside the range of adaptive, ecological behaviors and functions already present in the ancestral species. With narrowly adapted species, on the other hand, relatively less adaptive divergence is required for the fledgling descendant to become ecologically different enough from its ancestor that it stands a chance of survival.

It seems obvious that the greater the degree of adaptive change required for fledgling survival, the less likely that survival will be realized. It is simply easier to generate lesser, rather than greater, amounts of adaptive change per unit of evolutionary time. Put another way, narrowly adapted, ecologically specialized species have a greater a priori chance of producing successful offspring than their ecologically more generalized collateral kin. Ironically, phrasing the problem this way intersects the ultra-Darwinian perspective in an interesting way. The fundamental ultra-Darwinian credo sees organisms within populations (or even their genes, *fide* Richard Dawkins) locked in an interminable competitive struggle to leave more of their genes behind, to outreproduce one another.

I am not suggesting here that species compete with one another to create more descendant species. Far from it. It was Vrba's great contribution to point out that whatever is causing greater rates of speciation (as well as extinction) in lineages of specialized species, the cause lies rooted in aspects of the biology of the organisms themselves. That seems abundantly clear and certain to prove true. Whether one agrees with Vrba or with my preferred explanation, both arguments root the cause of higher speciation rates among specialized species firmly in the details of the biological properties of the organisms themselves.

It is *organisms* that are generalized or specialized. Saying that *species* are ecologically specialized or generalized is shorthand for the clumsier, but more accurate, characterization that it is organisms, not species, that are the ecological and adaptive generalists or specialists.

Thus Vrba saw relative rates of both speciation and extinction within related lineages as a fallout—a side effect—of the modes of adaptation of the component organisms. Vrba made it clear that ecological generalization versus specialization is but one example of a possibly greater range of such potential biases of speciation and extinction. However, so far no one has proposed any alternative to the spectrum of ecological adaptations as a cause of such a bias.

Vrba dubbed organism-engendered biases in speciation and extinction rates the "effect hypothesis." Ironically, she chose the name in recognition of George Williams's 1966 discussion of the intricacies of demonstrating that an organismic structure is a true adaptation fashioned by natural selection to perform a specific function, rather than an accidentally useful side effect of selection for some entirely different function. Vrba saw that Williams's point held true at the next higher level—where a lower level (organisms) could effect the fates of entities at the next higher level—that of species.

The plot has thickened. Without injecting arcane mechanisms into the dialogue, we naturalists have shown that recurrent patterns in evolutionary history are only explicable if we acknowledge the real existence of species. The properties and behaviors of organisms remain central to the evolutionary problem. But we are way past the point where we can accept simple extrapolation of linear natural selection within species as the way long-term "macroevolutionary" evolution occurs. We have had to rethink how the properties of organisms, and the workings of natural selection, fit into the vast sweep of evolutionary history.

Extinction and Speciation: The Ecological Nexus

Evolution, of course, doesn't occur in a vacuum. When I did my initial study on the *Phacops rana* trilobite lineage, I followed tradition and looked only at my trilobites. I ignored, except in passing, the 300-odd other species that lived in the same ancient Devonian seaways. I saw their fossils lying about, but only had time, and eyes, for my own quarry.

'Twas ever thus in evolutionary biology. Everyone focuses on a particular group, and such specialization quickly becomes forbidding. In the nineteenth century, a paleontologist could reasonably expect to become master of an entire fauna, or even a series of faunas, if one's career lasted long enough.

Not so any more. Paleontologists work hard to learn the anatomical and classificatory intricacies of a single group. Our charge is to understand that group's history, which demands attention to all species in that group regardless where they occur. It is the ecologist's job to know all the different species in a single area. The evolutionist, at least traditionally, looks at individual lineages, with their component species spread, living at disparate times, in far-flung places. But all that is about to change.

Evolutionists are not so naive as to imagine that their species evolve context-free. Natural selection, after all, is a matter of differential economic success biasing reproductive success. Relative economic success is a measure of how well an organism copes with its environment—the physical world of its habitat, plus all those other organisms with which it competes. Nils Stenseth and John Maynard Smith tried to apply Leigh Van Valen's "Red Queen" model—explicitly trying to incorporate a cross-genealogical ecological perspective into their evolutionary modeling.

We naturalists, too, have been busy looking at the intersection of the traditionally genealogical world of evolution with the highly charged world of ecology. And we have been turning up some amazing, and wholly unexpected, things.

Take the complex of ecosystems in which my Devonian trilobite lineage lived. We are back some 380 million years, in a shallow-water marine environment of the Devonian tropics. North America lay astride the equator in those days. I am currently engaged with University of Rochester paleontologist Carlton E. Brett in a massive study of all the species of the so-called Hamilton fauna—a group of some 300 species of hard-shelled marine invertebrates. The study has just begun; but we already know some fundamental and rather arresting facts about the Hamilton fauna.

The sediments forming the rocks of the Hamilton Group took some six million years to accumulate. In my *Phacops rana* lineage, I found two apparent instances of speciation, each involving only minor amounts of anatomical change. The central species in the lineup evolved near the very beginning of Hamilton time, and persisted to the very end. It is the paradigmatic instance of stasis. Nothing discernible happened to it throughout that six-million-year interval, except some minor proportional changes discernible only to my computer.

We suspect that the vast majority of the rest of the 300 or so species of Hamilton invertebrates tell exactly the same story. Indeed, I am convinced that once the study is complete we will find that the entire fauna is in stasis, with only minor fluctuations (such as my colleague Bruce Lieberman has already reported) breaking the nearly total monotonous grip of sameness.

While the results are still preliminary, it is already clear that the vast majority of Hamilton species were there right from the beginning, and last right up to the end. Indeed, the entire fauna appears rather abruptly, persists with great monotony (and much habitat tracking), and eventually disappears with the same abruptness with which it first appeared.

Nor is the Hamilton unique. Brett and his colleague Gordon Baird have documented ten such faunas in the Middle Paleozoic of the Appalachian region. Each lasts some five to seven million years. In each, from 70 to 85 percent of the individual species were there at the start and persist to the very end. Most appear to remain more or less static throughout their ranges. What is more, Brett and Baird calculate an average of only 20 percent of the species of any one fauna (for example, the Hamilton) are holdovers from the immediately preceding fauna. The rest of the preceding species disappear from the record, most of them presumed victims of extinction. Some of the new species appear to come in from elsewhere (many Hamilton species, for example, are invaders from what is now northern Africa and western

Europe). And some are newly evolved—from ancestors of the preceding fauna or from those living elsewhere.

What's going on? It's not just this or that individual species but entire regional marine faunas that remain incredibly stable; appearing, living, and disappearing in lockstep fashion. It's almost too much even for a confirmed punctuationally inclined naturalist like me to believe. But that's what the record seems to be telling us.

Where are those species coming from? Why do most of them live side by side for so long and then suddenly disappear? If we turn to a more thoroughly studied example, and one that took place more recently, it gives us a better chance to understand the environmental changes that apparently underlie these patterns.

Again we turn to Africa, and once again to Elisabeth Vrba's work documenting evolutionary patterns in fossil antelopes. Vrba finds much the same sort of pattern affecting the biota of eastern and southern Africa 2.8 million years ago as Brett and Baird see in the origin of Hamilton fauna: abruptly, most of the species living there simply disappear. Not just her antelopes, but all manner of plant and animal species of the East African ecosystems disappear. Right above this datum, a whole new cadre of species appears, with only a few (presumably all ecological generalists) of the older species making it through.

What caused this abrupt faunal turnover? Evidence is mounting that it was a pronounced event of global climate change. Paleoclimatologists have been hard at work for the past 20 years or so trying to understand the causes of the Ice Age— oscillatory drops in global temperatures, accompanied by the growth of massive continental glacial sheets and montane glaciers in the northern hemisphere. The earliest documented glacial advance of this most recent of the many ice ages that have occurred in geological history began some 1.65 million years ago—the official start of the Pleistocene Epoch. But paleoclimatologists have now detected an initial steep decline in global temperatures precisely 2.8 million years ago.

Vrba thinks this global climate change underlies the dramatic turnover in the African biota 2.8 million years ago. Essentially what occurred was a sudden, dramatic conversion of East African habitats from tropical woodland to more open savannahs, dotted by small woodlands. These are the sort of savannahs that still characterize the East African landscape today.

There are two distinct categories of events that can account both for the sudden disappearance and the consequent sudden appearance of all those species. One, of course, is habitat tracking. Some of the older (that is, pre–2.8 million year) species presumably were able to relocate where woodland habitats persisted. Similarly, some of the newly appearing species may have tracked the new, savannah-style habitat from elsewhere, where they had already adapted to life in such conditions.

Some of the disappearances apparently were true extinctions, while some of the new, abrupt appearances, by the same token, appear to represent newly evolved species. And that is a more difficult problem than has usually been acknowledged.

I have always been troubled by the traditional evolutionary supposition that evolution springs into action to fill empty niches. Why do species fragment and diversify, even after a major extinction event erases so many other species that had previously played all the roles in the older ecosystem? What, in other words, does economic opportunity (empty niches) have to do with the fragmentation of reproductive systems (speciation)? How would that work? The old model was never fully articulated. People seemed to think that new species just naturally pop up to take over the vacated roles left by their newly extinct forerunners. But it seems to be just a variant version of the old Darwinian-inspired notion that, if environments change, natural selection will, as a matter of course, mold organisms to fit the newly changed environmental conditions.

Speciation is a matter of fragmentation of preexisting species, thereby multiplying the number of species. If we think we understand how those species—those reproductive communities—become fragmented, we also see that the connection between speciation and adaptive evolutionary change is by no means straightforward. I could never see how empty niches would somehow induce species fragmentation. And yet, that is, at least superficially, what the record of such events seems to be telling us.

Once again, Elisabeth Vrba has made a simple and elegant suggestion that goes a long way towards a solution to this puzzle. Her turnover pulse hypothesis rests on one key observation: habitat disruption (and such it was in Africa 2.8 million years ago) has two basic effects. The first we readily see: rapid and profound habitat alteration causes species to disappear, whether by habitat tracking or by true extinction. On that we all agree.

But habitat disruption can also induce the appearance of new species—through habitat tracking as well, but also through speciation. Vrba points out that habitat disruption leads to disjunct populations, the very precondition needed for true species fragmentation. It is not empty niches that drives the evolution of new species. It is, instead, the environmental change itself—the very forces that drive some species out and others to extinction—that underlies a burst of speciation and brings others in from elsewhere. Climate change induces extinction *and* speciation simultaneously, along with habitat tracking. All components of the "turnover"—the abrupt disappearances of older species, and their rapid replacement by others—can be traced to a single climatic event. Pretty neat.

The turnover pulse model has the ring of truth to it, and the virtue of utter simplicity. T. H. Huxley reportedly said, "How extremely stupid not to have though of that," upon first hearing of the Darwin-Wallace notion of evolution through natural selection. I feel the same way about Elisabeth Vrba's turnover pulse idea.

But, much as I like it, the turnover model can't explain all patterns of rapid turnover. Sometimes the evolutionary response does not appear to be as immediate as Vrba's example of climate change and faunal turnover 2.8 million years ago. Especially in more globally massive extinction events, there is invariably a lag (sometimes millions of years) before substantial numbers of newly evolved species put in their appearance. The lag simply means that these later-evolving new species cannot be appearing as a simple response to the initial environmental change that caused the demise of their predecessors. So we need some additional theory.

That theory is, once again, ably supplied through the now familiar notion of species sorting. Once again, we must consider fledgling species. In a world suddenly devoid of a majority of the species that used to staff the ecosystems, we can imagine the probable fate of the few new species that might, in the course of time, appear through normal speciation. They would almost certainly have a far better chance of becoming established in a newly underpopulated world than would be the case in more "normal" times. If not quite a case of "anything goes," surely the exigencies normally facing fledgling species would be greatly relaxed after large-scale extinctions. Survival rates would be far higher than usual. It is the rate, not of speciation per se, but of successful speciation, that goes way up after a major extinction event.

Species sorting is an indispensable model for thinking about absolutely all evolutionary patterns involving more than one species. Even if my own penchant for biases in survival rate of fledgling species turns out to be less compelling than some other mechanism for species sorting, it is the patterns themselves that are most important. They demand explanation, and that is the job of evolutionary theory.

Missing in this section is any substantial mention of ultra-Darwinian positions on these issues. They are missing because the patterns are all paleontological in nature, and until very recently have not been brought to the attention of the evolutionary community at large. It should be clear, by now, that additional theory—as embodied in that general class of models we call "species sorting"—is not only justified, but absolutely mandatory. These patterns, so characteristic of the large-scale elements of life's history, are real and demand theoretical evolutionary explanation. It is equally abundantly clear that no manner of old-style, simplistic ultra-Darwinian extrapolation can be of any use in addressing these major patterns of evolutionary history.

Mass Extinctions and Evolutionary Response: Grand Scale Adaptation

Punctuated equilibria held the spotlight on paleontological center stage throughout the 1970s, and has remained an important focus of debate within broader evolutionary circles well into the 1990s. But another major subject of paleobiological concern—mass extinction—exploded in the early 1980s. Interest in extinctions was driven in large measure by the fascination over the Alvarez impact hypothesis for terminal Cretaceous extinctions 65 million years ago. Dinosaurs intrigue the scientific world every bit as much as they thrill the laity, and the possibility that a collision between the earth and an extraterrestrial object lay behind the demise of these beasts proved irresistible. Add to that the growing concern that the world's modern species currently face an episode of mass extinction of truly epic proportions—a truly millenial concern—and we have all the ingredients for a major research effort.

Extinctions, on scales ranging from relatively local disruptions and species loss on up through the five or six truly global mass extinctions that took away millions of species, are inherently ecological affairs. They are cross-genealogical, taking out unrelated species whose members populate the affected area's ecosystems. Yet there are profound evolutionary implications, as well. New ecosystems rise phoenix-like from the ashes of the old, fashioned from what species manage to survive and those that appear de novo. Paleontologist David M. Raup calculates that possibly as many as 96 percent of the world's species became extinct in the most monstrous extinction event that life has so far experienced: the Great Dying at the end of the Permian period, some 245 million years ago.

It is, of course, the evolutionary implications of mass extinctions that attract High Table attention. Once again we confront patterns in life's history that were well known to nineteenth-century naturalists, but have been generally ignored under the Darwinian evolutionary paradigm. Baron Georges Cuvier, French nobleman and outstanding anatomist, proclaimed that there had

been some 33 biological "revolutions sur la surface du globe"—
faunal turnovers that Cuvier saw as evidence of extinctions followed by separate creations. Such creationist interpretations,
coupled with the catastrophic, highly "punctuated" nature of
extinction events collided with Darwinian evolution, with its
strong predilection for gradualism.

By the mid–twentieth century, only a single prominent
paleontologist was devoting any time at all to documenting and
analyzing major extinction pulses. That was Norman D. Newell
of the American Museum of Natural History and Columbia University. He was mentor to myself, S. J. Gould, and a goodly number of like-minded fellow graduate students. We, at least, were
not surprised when attention suddenly turned in earnest to
extinction phenomena. And Gould, in particular, has not been
loathe to develop the evolutionary implications posed by extinction processes.

Demographers showed long ago that most people alive at
any one time do not have living descendants, say, 500 years later.
Put another way, all of us are descended from only a fraction of
all those people alive 500 years ago. The same holds for species.
The vast majority of species that have ever lived have not only
become extinct, they also have left no descendants.

But mass extinctions are something else again. Think of the
consequences if 96 percent of the world's species become extinct
over a relatively short interval (a half-million years according to
scenarios I personally prefer; a few years or at most decades
according to impact theorists). It takes upwards of five or even
ten million years for the world's ecosystems to regain a sem-
blance of normalcy after such events. Millions of new species
evolve—all descendants of a measly 4 percent of the previously
existing species. The world's genetic information passes through
an incredibly tight bottleneck if 96 percent of all species go
extinct in a single event.

The greater, more massive, more nearly global an extinction
event is, the greater the evolutionary consequences. More often
than not, at the boundaries between Carl Brett's mid-Paleozoic
Appalachian faunas, the extinction/evolution nexus remains at
the species level. Old species disappear, and new, often closely

related species appear and assume comparable roles in newly constituted replacement ecosystems. But in more widespread and devastating events, so many species become extinct that entire groups of species (taxa, such as genera, families, and even orders) will disappear.

As a general rule, the higher the rank of the taxa eliminated through extinction, the higher the rank of the taxa that replace them. When entire orders, for example, are wiped out, they tend to be replaced by other orders, as happened in the geological history of corals. Patterns of extinction and replacement of higher taxa are somewhat a bone of contention, as cladists point out that many of the taxa recognized by paleontologists are not, strictly speaking, genealogically pure. Even dinosaurs are not a pure group—if by "dinosaurs" we only mean terrestrial species, and forget to include birds. Birds are close relatives of theropods (carnivorous dinosaurs, including the famous *Tyrannosaurus rex*), and a paleontological cladist will tell you, correctly, that dinosaurs in a very real sense are not extinct—pointing to house sparrows flitting about outside the window.

Be that as it may, the general rule of congruence between levels of taxa killed off and those evolved in their stead holds. The corals of the Paleozoic (Order Rugosa) became extinct during the great episode of massive die-off 245 million years ago. Some five million years later, modern corals (Order Scleractinia) appeared. Scleractinian corals are more closely related to sea anemones, which lack a hard outer skeleton. It appears certain that modern corals were derived from naked anemones, and not from the Paleozoic rugosan corals, whose internal skeletal anatomy differs markedly from modern corals.

Thus we paleontologists are impressed first and foremost by the reinvention of the ecological wheel. The most dramatic effect of extinction on evolution is the subsequent production of species whose component organisms play analogous roles in nature to those wiped out in the extinction event. It sometimes seems, to me at least, imbued as I am with the patterns of utter stability in marine hard-shelled invertebrate faunas of the Paleozoic, that little or no evolution occurs unless and until an extinction event occurs to shake up entrenched ecosystems. Even though habitat disruption and consequent extinction inject a chaotic element into normally placid ecosystems, remarkably similar replacement ecosystems are normally rebuilt with newly evolved players.

Perhaps the most dramatic example of this principle lies in the great dinosaur-mammalian evolutionary interplay. Mammals and dinosaurs appeared at around the same time, in the Upper Triassic, some 210 million years ago. The dinosaurs quickly diversified into a prodigious array of species, playing a wide variety of ecological roles in terrestrial ecosystems. There were a number of different kinds of herbivores, and a smaller array of carnivorous and scavenging taxa to dine on the herbivores.

Mammals, in contrast, stayed small and rather generalized. Though arrayed in a number of higher taxa, mammals remained inconspicuous and were minor players in the Mesozoic ecological arena. Several times extinction knocked off large numbers of dinosaur taxa. Each time evolution produced more species, and the general complexion of Mesozoic life continued. Finally, though, 65 million years ago, whether dispatched with a bang through cometary impact or with the whimper of declining numbers in the face of global climate change, all terrestrial (non-avian) dinosaur species became extinct.

Then, and only then, did the mammals begin to radiate into the diverse array that we see today. (Actually, the initial mammalian radiation of the Paleocene quickly produced some very large herbivorous species, all of which became extinct. Early members of most of the mammalian orders that dominate today began to appear in earnest in the Eocene, following the extinction of the earlier, so-called archaic orders.)

The evolutionary moral of the story is obvious. Extinction is not a matter of the older, more primitive stocks giving way to superior, more highly evolved forms. The moral is especially clear if the deus ex machina extinction process involves collisions with rocks from outer space. Dinosaurs can hardly be blamed for such accidental events. No organisms are adapted to survive such calamities. Whatever the cause, it wasn't competing mammals that finally pushed terrestrial dinosaurs over the brink of extinction. Mammals were held in evolutionary check until after whatever did the dinosaurs in had done its job.

Evolutionary history, then, is deeply and richly contingent. Gone are the last vestiges of the idea that evolution inevitably and inexorably replaces the old and comparatively inferior with superior new models. Evolution, at least on a grand scale, is not forever tinkering, trying to come up with a better mousetrap.

It's the other way around: species, and the ecosystems that their component organisms staff, are tenacious. They "work" perfectly well and, once entrenched, are unlikely either to change or to be displaced by newly evolved taxa—unless and until extinction knocks ecosystems off their tracks. Then evolution breaks loose.

Or so it now seems. Richard Dawkins, reviewing Gould's *Wonderful Life* (1989), has tried to make a case that such contingency in evolutionary history is completely inconsistent with the evolutionary determinism embodied in "species selection." In *Wonderful Life*, Gould paints a vivid picture of what he calls the great "disparity" among marine arthropods living in Middle Cambrian times, some 520 million years ago (Cambrian dates are currently under revision). There are, he claims, far more different kinds (higher taxa) of arthropods preserved in the Burgess Shale than we see around us today. Evolution produced an initial burst of different arthropod body plans. Extinction has claimed all but a handful of them, leaving crustaceans, chelicerates (spiders, scorpions, and horseshoe crabs), and the great insect lineage. It could easily have gone another way had, say, the Burgess Shale crustaceans fallen in any of the subsequent major extinctions, and had some of the otherwise unfamiliar and downright bizarre early arthropods made it through.

That Dawkins sees inconsistency between this picture of historical contingency and the concept of species selection only underscores ultra-Darwinian reluctance, or sheer inability, to imagine evolutionary phenomena occurring at different scales and levels. For example, let us grant that some form of species selection (I would prefer "species sorting" for reasons already stated) has operated in addition to natural selection to orient the trend of increase in brain size in human evolution over the last four million years. If (as seems altogether possible) *Homo sapiens* becomes extinct in the future, marking the end of this four-million-year-old hominid line, the implications for continued existence—and possible further evolution—of virtually all other species would be considerable. It is obvious, in other words, that the continued existence of many of the world's species is contingent on our own fate. If we drop out, they stand a good chance of continued survival. If we persist along our current path of mass destruction of the world's ecosystems, many species will die—perhaps paving the way for future evolution (especially if we ourselves, sometime further down the road, also become extinct).

There are clearly levels of severity in the interplay between extinction and evolution. Regional extinction can play an important role in species sorting sequences. This was evidently the case 2.8 million years ago when habitat disruption apparently caused the extinction of *Australopithecus africanus*, leading to the appearance of species of *Paranthropus* (an evolutionary sideline—one which produced no discernible increase in brain size) and *Homo habilis* (our own early progenitor, with a marked increase in brain size over *A. africanus*).

The real point is that evolution can only work on what survives. Older models of species selection invoke a form of differential survival reflecting greater success of some species within a lineage over others—in, as we have seen, a rough analogy with natural selection. Relative survival of species, in this model, depends on the relative success of the adaptations of organisms within the various species of a clade.

Most forms of the historical contingency argument take another tack. Relative survival may have nothing to do with the relative "goodness" of an organism's adaptations, but simply reflect bad luck. It was bad luck when the comets struck (that is, if they did), seriously impeding photosynthesis, seriously compromising both marine and terrestrial ecosystems, and leading to the demise of many species. The dinosaurs became extinct, not because they were poorly adapted, but because the systems in which they lived were so severely altered. It was just bad luck, according to this mode of thought, that the dinosaurs succumbed.

Thus there is no more inconsistency between the determinism embodied in species sorting models and the chanciness of historical contingency models than in the twin factors universally conceded to underlie change in gene frequencies within populations: natural selection (statistically deterministic) and genetic drift (with its strong element of chance). Dawkins's critique of the themes of *Wonderful Life* falls apart, hinging as it does far more on snide rhetoric than substance.

There is a lot more to the contingency argument. Gould says, in effect, that if we were somehow able to start the evolutionary story once again around Burgess Shale times, life today would be greatly different. The chances that *Homo sapiens* would be here are vanishingly small.

But there are repeated adaptive themes that show up again and again in evolutionary history, adding a deterministic element to the mix. Whatever the taxa involved, the basic complexion of ecosystems—say, terrestrial temperate woodlands—is repeatedly assembled, even if the plant and animal species have been continually churned since life colonized land 400 million years ago. That leaves us with a final consideration: What can we say about the appearance of the truly new adaptations, such as the origins of bird flight, in evolutionary history?

Most paleontological naturalists have lately been emphasizing the utter dependence of evolution on extinction. Yet it is quite true that most reactive bursts of evolution following extinction events are consumed with reinventing the ecological wheel. Scleractinian corals, true enough, constituted a major new taxon (Subclass Scleractinia, Class Anthozoa, Phylum Coelenterata) when they appeared in the Triassic. But scleractinians were merely replacing the extinct rugosans and other reef-building calcareous organisms of the Paleozoic. In that sense, scleractinian evolution does not address the issue of the "truly new" in adaptive evolution.

Closer to the mark are instances where major new habitats are invaded for the first time, as in the occupation of the land—

when true vascular plants and several independent lineages of air-breathing arthropods, mollusks, and vertebrates first appeared. Even the relatively sudden burst of evolution that established the major phyla of multicellular animal life at the "base of the Cambrian" 535 million years ago fits in here. It seems most likely that that event was triggered by the crucial environmental change of a rise in the partial pressure of oxygen in seawater to a critical level high enough to support large, multicellular oxygen-dependent animals.

There are several models currently being debated to explain why so many different phyla appeared in the Cambrian, and why no phyla have appeared subsequently. No phyla are definitely known to have become extinct either, so the reason no new phyla have arisen since the Cambrian proliferation might simply reflect the general pattern we have already seen. Evolution produces new taxa generally ranked at about the same level as the taxa that disappear in extinction events.

Evolutionary radiations following the invasion of wholly new environmental realms (such as the mid-Paleozoic invasion of the terrestrial realm) are inherently like evolutionary recoveries after extinction events. In a sense, anything goes in such a situation—meaning that survival of fledgling species is way up: species sorting once again. In his recent book, biologist Stuart Kauffman, who frankly seeks a "physics of biology," has produced mathematical models of the Cambrian explosion, concluding that "poorly fit, multicellular organisms [could have rapidly explored] a large diversity of improved alternative basic morphologies, thereby establishing phyla." Kauffman actually sees a contrast between the Cambrian explosion and the relative "quiescence" following the Permian extinction. No new phyla were produced after this because (according to Kauffman) the developmental pathways of the survivors, who went on to found the subsequent Mesozoic ecosystems, were too-long established, locked-in, and inflexible.

In any case, whatever the differences between invasion of wholly new territory (metazoan radiation in the sea, invasion of plants and animals of the land in the Silurian and Devonian periods) and the recovery and reestablishment of ecosystems after an episode of mass extinction might be, the notion of a high rate of increase of fledgling species—species sorting—is common to explanations of both sorts of situations.

Not all evolutionarily inclined paleontologists buy the argument that evolution depends so heavily on extinction. Leigh Van Valen and colleagues have openly disputed this generalization. Van Valen, recall, trained simultaneously under geneticist Theodosius Dobzhansky and paleontologist George Gaylord Simpson. He is the author, as well, of the Red Queen hypothesis, and a number of other intriguing evolutionary suggestions. By no means classifiable as one of the naturalist camp, Van Valen is no knee-jerk ultra-Darwinian either. He fits better on that side, but he is more sui generis than a lot of evolutionary biologists can claim to be.

Van Valen has pointed to a number of examples where significant evolutionary adaptive innovation seems definitely *not* to be linked to either an immediately preceding extinction event or to the overt and demonstrable invasion of new environments. Teleost fish and flowering plants both appear in the Lower Cretaceous. By the end of Cretaceous times, both had radiated explosively. Neither group was badly damaged by the end-of-Cretaceous extinctions, and both survive to dominate the ecological landscape today. Just about every bony fish species on earth today is a teleost, and the overwhelming majority of land plants are angiosperms.

Teleosts have major adaptive innovations—most important of which, perhaps, is the freeing of the jaw apparatus, allowing a wide range of feeding behaviors to develop. The adaptive innovation of angiosperms is reproductive. Their name means "covered seed" (as opposed to the "naked" seeds of gymnosperms, such as the pines and their relatives). Angiosperms seem able to live and reproduce in a wider range of habitats than gymnosperms, which have been relegated to a progressively narrower range of habitats over the past 100 million years.

Van Valen and his colleagues are perfectly right: not all major episodes of adaptive change need follow a major extinction event, or even an invasion of new habitat. But all such events do involve the proliferation of species. George Gaylord

Simpson pointed out years ago that many such instances hinge on the development of what he called "key innovations"—adaptive modifications that seem to trigger evolutionary explosions.

Under such circumstances, speciation rates, meaning the rate of appearance of *successful* species, shoot way up. Natural selection fashions adaptive change; speciation conserves the change and, if the novelty is truly useful, a chain reaction of successful speciations ensues. Species sorting very definitely plays its accustomed role alongside natural selection in such instances. Adaptive change occurs without prior extinction, but not independent of the process of speciation.

All instances of adaptive radiation demand a combination of adaptive change in the context of speciation and species sorting as the clade proliferates into a large array of species. Apart from a few treatments (notably by ornithologist Walter J. Bock), not many evolutionists have explicitly integrated models of speciation with their adaptive scenarios of natural selection producing patterns of adaptive change in evolution. It is high time that we did.

Species Sorting in the Modern World: Latitudinal Diversity Gradients

Ultra-Darwinians, of course, have no direct access to paleontological data, partly explaining their inability to understand species sorting and its relevance to evolutionary theory. But species exist in the modern world, and exhibit patterns that clearly arise from differential births and deaths of species. Adaptive radiations are obviously one such pattern. Latitudinal diversity gradients are another.

Biologists have long wondered why there are so many more different kinds of organisms—animals, plants, fungi, and microorganisms—in the tropics than in the higher latitudes. The tundra regions of the far north support few species, and often huge populations—such as vast herds of caribou. The tropics, in contrast, seem jammed with many different animal and plant species, but commonly harbor very small numbers of any one species in any single place. There are often hundreds of plant species present on a single hectare of tropical rain forest, but only a handful are really abundant. Most species are represented by only one or two individuals.

The pattern has both its evolutionary (phylogenetic) and ecological aspects. Frequently there are more species within a higher taxon as one samples from the higher latitudes into the tropics. There are far more species of wrens, tanagers, and flycatchers the farther south one travels from North America to the tropical regions of South America. But there are definite exceptions: There are far more species of penguins in the antarctic than there are in the tropics (the Galapagos penguin is the only species to sit astride the equator).

The pattern is distinctly ecological as well. There are simply more species comprising the ecosystems of the tropics than in most higher latitude systems. But, here again, there are exceptions. A tropical desert, for example, will not support as many species as a temperate woodland forest no matter how close to the equator it might be.

Ecologists tend to tackle the problem in strictly functional terms. There is more solar energy in the lower latitudes, hence productivity might be expected to be higher than in the higher latitudes. There might be more ways to make a living in the tropics—more "niches"—though we run into problems of circularity with such rhetoric, as niches are defined and recognized by the species occupying them. (Strictly speaking, local populations, rather then entire species, play roles in local ecosystems, and thus have niches. The dialogue linking ecology with evolutionary theory has been beset by such casual and often erroneous usage of terms.)

Evolutionists, in contrast and as might be expected, have stressed historical explanations for latitudinal diversity gradients. The tropics have often been considered a safe haven, a cushy environment, almost a museum of living diversity. Eurasian and North American counterparts of the threatened but still-extant African megafauna—buffalo, antelope, elephants, rhinos, hippos, and lions and other carnivores—largely disappeared in a wave of extinction toward the end of the last glacial pulse (at the same time that *Homo sapiens* was expanding its range throughout the globe).

On the other hand, evolutionists have also postulated the tropics as more of a nursery than a museum—a laboratory of evolutionary diversification where speciation rates are far higher than in the higher latitudes. Such concepts dovetail more closely with traditional ecological explanations, and correspond most closely with the species sorting hypothesis.

Indeed, a species sorting approach to the problem of latitudinal diversity gradients draws heavily on recent work in ecological theory—most notably by George Stevens of the University of Arizona. Stevens himself drew on the work of biogeographer Eduardo H. Rapoport, who compiled areal distributions for subspecies of 197 species of mammals from North and Central America. Rapoport's data revealed that, as a general rule, species in higher latitudes tend to occupy greater latitudinal ranges than their lower-latitude counterparts—a pattern that Stevens dubbed "Rapoport's rule."

Stevens compiled additional data, and claims the pattern holds for all manner of organisms, except those that hibernate or migrate. For an explanation of why species would have greater latitudinal ranges in the higher than lower latitudes, Stevens turned to the by-now familiar distinction between generalists and specialists. There is simply greater daily and seasonal fluctuations in climate—most obviously, in ambient temperature—the further away from the tropics a species lives. Simply because all species living in the higher latitudes must be sufficiently physiologically generalized to withstand these fluctuations, they will be able to live over broader areas.

Tropical species do not face such pronounced climatic fluctuations (though there are often pronounced seasonal variations in rainfall). Stevens thinks this simple fact is part of the puzzle: the greater equability of the tropics allows species to be more specialized, focusing more narrowly on habitat parameters. In ecological parlance, species in the tropics tend to perceive their habitats in a more fine-grained manner. In a very real sense, Stevens is providing a variant version of the claim that there are more "niches" in the tropics.

Thus the gross distinction between high- and low-latitude species mirrors the contrast that Elisabeth Vrba drew between impalas (the single species generalist) and the many specialized species of the wildebeest clade. Her distinction was between two closely related antelope lineages living in the same area. In the latitude story, we contrast unrelated lineages living in different areas, but the net effect is the same: there are more species of specialists than there are of generalists. Indeed, the parallel has even more force, because some ecologists have maintained that the latitudinal disparity in species number is offset by numbers of individuals. The total biomass, in other words, in higher latitudes may be the same as in the tropics, with vastly more individuals of fewer species roughly equaling the fewer individuals of the many more species that live in lower latitudes. Recall that Vrba and Greenacre found precisely the same in their antelope census data. There are about as many impala individuals as there are individuals of all the other species of wildebeests, hartebeests, and topis living in any one area.

Stevens supplemented Rapoport's rule with another ecological observation. All species have a finite range in nature (even *Homo sapiens*) and all are limited by habitat distribution. As we have already seen, the choicest habitats vis à vis the adaptive requirements of a species are around the central regions of a species' distribution. Conditions are less favorable at the periphery. Often, as Stevens points out, organisms living at the far edges of a species' normal range are living in such suboptimal habitat that reproduction is sharply curtailed. Such physiologically weakened organisms often find themselves living far from potential mates. Stevens (following ecologist Dan Janzen) calls them the "living dead": they are clinging to a precarious existence, but they are not reproducing.

Stevens hypothesizes that the tropics are far more full of the "living dead" than are higher latitude regions. Species may seem to be widely distributed in the tropics, but many of them are represented by very few individuals. Stevens speculates that many of these single-individual outposts are far-flung colonists too far removed from their fellow conspecifics to be effective members of the breeding population. The effective range of species in the tropics is, on average, even smaller than first meets the eye.

The tropics are no safe haven, no living museum for species—in spite of the persistence of the African charismatic megafauna for an additional 10,000 years. If anything, extinction rates are greater in the tropics than in the higher latitudes. Steven Stanley points out that the tropics are always the hardest hit during mass extinctions, further suggesting that tropical species are, on average, more narrowly focused adaptively than their counterparts in the higher latitudes.

One additional observation seems pertinent as well: related species in the tropics are often quite similar to one another. As biologist John C. Kricher points out in his *A Neotropical Companion* (1989, p. 148), "perusal of any of the current guides to birds for areas in South America reveals almost extreme similarity among species within certain groups..." This Kricher presumes to imply constitutes "evidence that speciation has probably been relatively recent." Indeed, inspection of Guy Tudor's color plates, for example, in *Birds of Venezuela* by de Schaunsee and Phelps, does reveal such extreme similarity, especially among antbirds, woodcreepers, and tyrant flycatchers. But note the gradualist assumption underlying Kricher's conclusion that such extreme similarity bespeaks recent speciation. What the tropics seem to hold, instead, is more speciation, with often less anatomical change arising between species—a pattern we have already encountered in lineages of adaptively specialized species.

Thus we return to the familiar model, species sorting, to explain disparities in species diversity between generalists and specialists in closely related evolutionary clades. Organisms in the tropics are released from the necessity of coping with huge daily and seasonal climatic fluctuations. They can specialize, and perceive minor variations of habitat, nuances less readily perceptible to higher latitude species. It takes less adaptive change at speciation for a newly evolved species to establish ecological distance from its ancestor. Once again, it's not speciation rates per

se, but the rate of successful speciation that is higher in the tropics. This also seems to be the case for specialists versus generalists within monophyletic clades.

Applications of species sorting models to latitudinal diversity gradients are obviously still in rudimentary form. We need
to know much more about Rapoport's rule—and if there really
are lots more living dead in the tropics than in the higher latitudes. We need to find out if there really are lots more sibling
(virtually identical) species in the tropics than in higher latitudes—an obvious prediction if the species sorting explanation
holds water.

But there is enough to the story to make it quite plain that
such models are, in principle, appropriate grist for the explana-
tory mill. It isn't just paleontological pattern that demands such
higher-level explanation. The modern world around us, filled as
it is with ecosystems and species, is constructed in such a way
that the need for this particular kind of theory should be obvious
to all. We really can do better than mouthing ultra-Darwinian
gradualist assumptions to explain ecological and evolutionary
patterns in all our large-scale biological data—be they drawn
from the deep past or strictly from the modern world around us.

6

Approaching Complexity

Evolution in the Real World

Geneticists, naturally enough, are interested in genes. We pale-ontologists have our fossils—partial remains of long-dead ani-mals. What *we* see are patterns of stasis and change in heritable anatomical features. So we paleontologists automatically know that genes must be important to the story. But we also think that the prodigious spans of time and space over which evolution proceeds are equally vital to the story. So are such larger-scale systems as species and ecosystems. These points are less obvi-ously clear to an ultra-Darwinian focused on genes: there is something asymmetrical happening here. Why is it easier to see the existence and importance of relatively small-scale entities, and so much more difficult to shift our gaze upward to embrace larger-scale entities and processes?

For one thing, we humans are virtually impelled to dig deeper, to analyze—literally to take apart—a system to see how it works. That means concentrating on smaller components, which we automatically assume must exist. In evolutionary biol-ogy, heritability demanded an explanation. The inferred and abundantly confirmed existence of genes and their workings are central to such an explanation. Theodosius Dobzhansky argued eloquently, though, that the dynamics of genetic change in populations play in a larger-scale ballpark than the workings of physiological genetics. Genes are copied (sometimes with errors: mutations) and are transcribed within the cells of organisms, while the processes of natural selection and genetic drift operate in a larger-scale arena involving many different, interbreeding organisms. It was Dobzhansky, building on the pioneering efforts of Ronald Fisher, J. B. S. Haldane, and Sewall Wright, who established the importance of the population over and above the workings of within-organism genetic processes. Dobzhansky provided the true nexus between the Darwinian vision of evolution and the newly derived theory of the gene.

More recently, in the ultra-Darwinian era, we have seen something of a reversion to the turn-of-the-century attitude that genes are not just important to understanding evolution, but that they are just about the only "things" we need really consider to understand the evolutionary process. Not that Richard Dawkins's notion of the "selfish gene" is truly a reversion to turn-of-the-century genetics. Back then, discovery of the elements of heredity prompted some enthusiasts to proclaim the death of Darwinism, especially the idea of natural selection. All that is required for a theory of evolution, it was thought, is a theory of inheritance.

Dawkins, of course, says something different. Genes do not exist in a vacuum. They reside in organisms, and organisms are parts of breeding populations. For that matter, Dawkins readily concedes that populations are sometimes complexly organized into social systems, and integrated with other populations to form ecological communities or ecosystems.

But organisms, to Dawkins, are merely the vehicles for the genetic material. The genes call the shots; they vie with one another to leave more copies of themselves to the next generation. The modern view of natural selection sees organisms competing with one another to leave more copies of their genes behind. This is the view embodied in the neo-Darwinian synthesis, and still the basic concept of natural selection in most evolutionary discourse today. Dawkins merely took the notion and pushed it a step further: it is not organisms, but rather their genes, that compete for "reproductive success." Genes need organisms to house them, to nurture them, to provide the energy and physical apparatus for them to function.

Genes, of course, as segments of DNA (and RNA) strands, are ephemeral—less stable and long-lasting even than organisms. What is at stake here is not the physical corporality of genes, but rather the information they contain—a point George Williams has been at pains to emphasize over the last twenty years. Dawkins sees competition between differing bits of information: between alternative forms on two paired chromosomes (alleles), or between alleles in different organisms within a population. As such, Dawkins's notions have nothing to do with turn-of-the-century ideas that the processes of heredity are both necessary and entirely sufficient to explain evolution. What the two views do share, however, is an insistence that what goes on in the world of genes determines evolutionary history.

Dawkins's "selfish genes" gambit posits, in effect, that genetic information, viewed as the instructions for constructing an organism, is more important than the system it builds—the organism itself. Evolution is strictly competition among genes for representation in the next generation. This is the ultimate reductionist scenario yet concocted in evolutionary biology.

Looking to smaller component parts of a system does seem to come naturally. But it also begs the question: how are we to treat larger-scale systems? Here, ultra-Darwinians and naturalists completely disagree. Ultra-Darwinians see large-scale systems as simple epiphenomena of lower-level processes. Competition for reproductive success leads to the formation of species, ecosystems, and social systems, but these systems themselves play no active role in the evolutionary process. I see the world very differently: large-scale systems arise from organismic economic and reproductive behaviors; whether or not organisms (or their genes) are in competition has nothing formally to do with the formation of such systems; and such systems themselves are vital aspects of the evolutionary process and of evolutionary history.

Ours is a hierarchical perspective. As such it dovetails nicely with parallel developments over the past few decades in other sciences. There are even connections with chaos theory. Ultra-Darwinians, in striking contrast, seem hell-bent on establishing a highly reductionist, old-fashioned physics model of simple lower-level determinism. The two approaches differ radically in the way they approach biological complexity. Reductionism is losing its grip elsewhere in science, and the time has come for us to reverse the trend and take another tack in evolutionary biology. That's where hierarchy comes in.

Approaching Complex Systems

In 1962, Nobel Prize–winning economist Herbert Simon delivered an address to the American Philosophical Society on "The Architecture of Complexity." Simon's discussion focused on the hierarchical structure of complex systems. It has become a landmark in the analysis of the complexity seen in a prodigiously diverse array of physical, chemical, biological, and social systems.

Hierarchy theory recognizes that complex systems are typically composed of smaller units which are themselves composed of still smaller entities. At each level, the entities are stable. Chaos theory, well described in James Gleick's *Chaos*, is itself an aspect of hierarchy theory, focusing on the often chaotic behavior of subsystems within an apparently stable larger entity.

Simon himself was primarily concerned with stability. In his 1962 paper, he tells the story of two watchmakers, Hora and Tempus, whose exquisite watches were in high demand. Their phones rang constantly with new orders. But, while Hora's business prospered, Tempus soon found himself out of work. It seems that Hora and Tempus had radically different approaches to assembling their 1,000-piece watches. While Tempus put his watches together piece-by-piece, Hora made ten-piece subcomponents that could be assembled later into a completed watch. Every time the phone rang, work was interrupted and the system under assembly would fall apart. This was of little consequence to Hora's work on subcomponents, but was disastrous to Tempus, whose entire watch would collapse even if he had 995 parts already in place. The more popular his work, the more the phone would ring, and the more difficult it became for him to finish a single watch!

That the universe consists of parts and wholes, and that parts can be seen as wholes themselves composed as parts, is intuitively obvious to everyone. What kid hasn't written the personalized equivalent of "Janet Smith, Sayreville, New Jersey, United States, North America, Planet Earth, the Solar System, Milky Way Galaxy, the Universe"? Then, too, everyone knows that bodies are composed of organ systems, which are composed of tissues, themselves made of cells composed of molecules, which are in turn composed of atoms. As Simon himself wrote in 1962, "only a couple of generations ago, the atoms themselves were elementary particles; today, to the nuclear physicist they are complex systems"—that is, composed of other, smaller things we now call "elementary particles."

Much of the history of modern science has been an exploration of progressively smaller bits and pieces of the material universe. Anton van Leeuwenhoek's microscope revealed the existence of cells in the 1670s. Advances in light microscopy improved resolution of tiny cellular constituents, and electron microscopy takes us right down to the molecular and atomic levels. Cloud chambers and mass accelerators have taken physicists down to the world of subatomic particles.

Consider atoms. As Simon says, we now see them as complex systems, rather than the least, indivisible building blocks of matter. But we still think of atoms as real entities, even though we realize that neutrons and protons, constituents of atomic nuclei, are themselves composed of smaller particles. With a blurry cloud of electrons swirling around the nucleus, we now understand atoms to be mostly empty space. Yet we persist in thinking of them, somehow, as real, concrete entities.

Looking down at small entities, such as the organs, tissues, cells, molecules, and atoms that compose our very own bodies, we seem to remain comfortable thinking of smaller, constituent entities as somehow "real." Not so when we look upwards, to larger entities. There is a problem of not seeing the forest for the trees. Ecologists are forever troubled with the lack of sharp boundaries setting off one ecosystem from another. An owl leaves a woodlot for an open field, and somehow the reality of the different systems—the woodlot and the field—seems in doubt. There is no analogue of the Hubble telescope in biology: but we should remember that, until the early decades of the twentieth century, astronomers thought that the Milky Way constituted the entire universe. Only when astronomers trained powerful telescopes on far-distant objects, revealing their spiral structure so like the Milky Way, did it become obvious that there are many other galaxies out there in the universe.

Only great distance, reducing galaxies to pinpoints of light so that telescopes must act like microscopes to bring the objects into view, enables us to see galaxies as real entities. No one doubts the reality of galactic systems. But I doubt that we would have ever concluded that the Milky Way is a "real entity" had we not discovered its kindred objects—so huge, yet reduced to such small images through such great distances.

Being part of a system seems to make it difficult to see that system as a concrete entity. Boundaries are sloppily defined, and most of the large-scale "entity" consists of open space—like a forest, a solar system, or a galaxy. We think of ourselves as most certainly real, with our outer tissue layers providing a sharply defined boundary between us and the rest of the material universe. But it is all a question of scale. Viewed on a microscopic level, our exterior surfaces are continually in flux, with cells sloughing off constantly while gases emanate from every pore. Could the mites that live on our skin think, there is no way they'd see our epidermis as a sharp boundary between some large-scale system and the rest of the universe.

The fossil record tends to condense large-scale biological systems, much the way intergalactic distances shrink the apparent sizes of galaxies, rendering them comprehensible as objects. A reasonably complete 1,000-foot sequence of sedimentary rock may desultorily record five million years of time, recording the entire history of a species from its first appearance, through samplings of its history, right up to the time of its eventual demise. It takes no time whatever to walk through 1,000 feet of sedimentary rock, sampling the entire five-million-year history of a species. Condensation of five million years of history into 1,000 feet of rock tends to make a paleontologist think of large-scale biological systems—species, to be sure, but also ecosystems—as real, concrete entities. Species have births, deaths, and eventual demises.

Hence the appeal of the "species as individuals" thesis articulated by Michael Ghiselin, and elaborated by David Hull in the 1970s. Stasis, as we saw in chapter 3, only enhances the view that species are real entities, making species inherently easier to recognize throughout their long histories. Ghiselin and Hull adopted their position for the very opposite reason. Species are traditionally expected to change radically during the course of their existence, so that they are "individuals" who, like individual humans, remain the "same" entity though they change with age. Ghiselin and Hull were contrasting their view of species as "individuals" with the more traditional notion that species are "classes," consisting of individual organisms that all share a set of defining properties.

Ghiselin's analogy was with atomic elements. Gold, for example, is a class, a category of atoms, all of whose members have the same atomic number: 79. Any atom with atomic number 79, occurring anywhere in the universe, and irrespective of where and when it was formed, is by definition an atom of gold. Each atom is an individual, but gold itself is a class.

Two traditions in biology reinforce the notion that species—indeed, all taxa of whatever rank—are classes, like gold. The first is the practice of taxonomy (systematics), beleaguered throughout its history with the reputation of being an arbitrary human exercise of classification, of sorting organisms into groups according to some human-defined sharing of traits. It is noteworthy that the history of systematic biology has been a story of discovering such arbitrary and erroneous groupings, and placing groups where they really belong. This, Darwin taught us, is based on "propinquity of descent," meaning phylogenetic relationship. We have long since stopped thinking of whales as fishes, and systematics continues to seek out comparable (if more subtle) mistakes. Taxa are real, the products of the evolutionary process.

But it was Darwin who, in seeing the very notion of evolution as antithetical to the idea of species as stable systems, helped perpetuate the notion that species are, in an important sense, not real. Darwin, and most of his successors, saw species as inevitably evolving themselves out of existence. Even Mayr and Dobzhansky, who labored so mightily for species to be seen as real, perpetuated the view that, once established, species inevitably accrue so much change that the biologist is forced to chop up the evolving lineage, recognizing two or more species within the confines of that single, evolving lineage. In that view, species are real only in the sense that they are part of a temporal lineage "reproductively isolated" from other lineages. But their vertical boundaries are vague, and are arbitrarily assigned by the biologist—the genesis of Walter Bock's comment (quoted in chapter 3) that the biological species concept has no temporal dimension.

We naturalists, particularly those of us who have spent our careers working with the fossil record, are completely convinced that species are real. Species are "spatiotemporally bounded historical entities." They have beginnings, histories, and ends.

Organisms, then, are parts of species. More accurately, organisms are parts of populations; populations are parts of species. Species, for that matter, are parts of larger-scale taxa.

And here is where Simon's parable of Tempus and Hora fits in, and where current interest in chaos theory dovetails with our naturalist, hierarchical perspective in evolutionary biology. Gene frequencies are constantly changing, through selection and drift, in each generation, in each and every population within a species. As Sewall Wright pointed out so long ago, each deme (semi-isolated population within a species) is undergoing its own, semi-independent evolutionary history. Indeed, it is that chaotic melange of semi-independent diversification within a species that adds up to little or no net change within a species. How can a far-flung species, with so much going on within each of its demes, possibly exhibit a single, directional, unified linear pattern of evolutionary change? It was Wright, as we saw in chapter 3, who gave us a strong explanation for stasis. All that evolutionary ferment and foment within local demes is mostly canceled out as populations come and go.

The implacable stability of species in the face of all that genetic ferment is a marvelous demonstration that large-scale systems exhibit behaviors that do not mirror exactly the events and processes taking place among their component parts. There are a number of perfect analogies ensconced in the canon of neo-Darwinian population genetics. For example, it has been known at least since the 1930s that the rate of mutations within a population has no direct relation to the rate of adaptive change in that population. Adaptive evolution depends ultimately on mutation as its source of variation, but generation-by-generation adaptive evolution is a function of natural selection acting on variation already present, and is not directly dependent on mutation rate.

Likewise, there is no correlation between time elapsed between generations and the observed rate of evolution within lineages. For example, it might be supposed that rats will evolve a lot faster than elephants, there being so many more generations of rats exposed to natural selection than elephants per unit time. Not so: elephants seem to have changed far more than rats in the past 50 million years. Simple lower-level processes cannot be extrapolated to yield long-term results.

Events and processes acting at any one level cannot possibly explain all phenomena at higher levels. That is why it has become so important to understand the part-whole relationships, the hierarchical structure, of biological systems. We cannot simply look at the dynamics of natural selection within a single population and assume that this alone is all there is to the evolutionary process: to the formation of species, to the generation of adaptive change that is incorporated into the phylogenetic mainstream, and to the very existence of larger-scale entities. It is those very entities—those stable entities that Simon stressed in his Hora-Tempus parable—that loom as critical to a deeper understanding, not just of the evolutionary history of biological systems, but of their very structure and function. We need to examine more deeply the nature of large-scale biological systems.

Large-Scale Biological Systems

Biologist Stanley N. Salthe, a herpetologist by trade, became known in evolutionary circles through his well-received textbook on general evolutionary biology published in the 1960s. I myself encountered Salthe's work somewhat later through his intriguingly, albeit confusingly, entitled paper, "Problems of macroevolution (molecular evolution, phenotype definition, and canalization) as seen from a hierarchical viewpoint." Salthe's paper was published in 1975 in *American Zoologist*, but I didn't stumble upon it until I found myself wrestling with biological hierarchies in an evolutionary context a few years later.

Issues of species sorting and selection automatically suggest a hierarchical organization to biological systems, and by the late 1970s, we naturalists had already begun using the term "hierarchy" freely. The sort of hierarchy we had in mind was composed of the traditional elements of evolutionary discourse going all the way back to Darwin: genes (a twentieth-century addition), organisms, populations, species, and higher taxa.

But Stan Salthe had something very different in mind. His title said "macroevolution" and "hierarchy," but his discussion was all about proteins, organisms, and ecosystems. This was interesting: here we had a paper supposedly about evolution that said next to nothing about the kinds of entities traditionally found in evolutionary discussions. Salthe was far more concerned with functional systems—systems where energy flowed from part to part. Salthe was talking evolution, but writing about matter-energy transfer processes: the sort of thing that happens when a fox eats a rabbit.

I told him I was baffled by his approach. We needed to link the two systems in some satisfying way. We needed to integrate them into a coherent hierarchical description of large-scale

natural systems that would embrace both the traditional genes–
organisms–populations–species–higher taxa elements of traditional evolutionary concern *and* the proteins–organisms–ecosystems that Salthe emphasized.

I gave it a shot in the early 1980s, but by the time the paper
appeared in print (in 1982), I already knew it was wrong. I simply tried to jam the two systems together. Biological nature is
arranged hierarchically, I claimed, and follows the form: genes
are parts of organisms, which are parts of populations, which
themselves are parts of species. I then inserted "ecosystems"
above species, and then wrote "higher taxa" as the last term. I
was claiming, in effect, that species are parts of ecosystems—a
common enough mistake in those days (and still to be found in
evolutionary circles). But I was also apparently saying that ecosystems are somehow parts of higher taxa, which appears in
retrospect, a scant decade later, utterly bizarre. Ecosystems are
composed of parts of many different, unrelated species, whereas
higher taxa are composed strictly of closely related species.

I knew my abortive, integrated hierarchical system was
wrong because, by 1982, I had begun an intense collaboration
with Stan Salthe. I had persuaded him that genes, species, and
higher taxa, as traditional elements of evolutionary biology,
existed and had to be accounted for—and themselves formed a
hierarchical system. He, in turn, persuaded me that the more blatantly ecological hierarchy that he had described does in fact
exist and is directly relevant to evolution. After all, most adaptations are concerned with issues of matter-energy transfer: the
wide spectrum of mammalian teeth reflect myriad adaptations
for eating different foodstuffs. Most adaptations, in other words,
are *economic* in nature—as we ended up terming them. So we
wrestled with how the two systems connect.

The first victory was to see that the two systems do exist
and are largely independent of one another. They don't go
together as one seamless, integrated system as I had originally
and artificially tried to fashion. Independence: that was the key.
Once we realized that there are two separate hierarchies of biological systems in the real world, we were free to look for connections between the two systems.

Salthe and I coauthored a manuscript detailing the two parallel hierarchies: the *economic* (or ecological) hierarchy that Salthe had first written about, and the *genealogical* hierarchy that I was and still am so devotedly attached to (the system that springs directly from traditional evolutionary theory). We sent our paper to the journal *Evolution*—which seemed the logical place. It was rejected by return mail, with the editorial comment that *Evolution* publishes papers on experimental and theoretical evolutionary biology, and not on philosophy.

Our savior turned out to be none other than Richard Dawkins! Dawkins had written me, soliciting manuscripts for a new journal to be published once a year in book format entitled *Oxford Surveys in Evolutionary Biology*. I wrote back, telling him about our paper, and suggesting that he would not like it: after all, it was about the existence of hierarchies, and the importance of large-scale systems to a complete understanding of evolutionary biology. Dawkins wrote back asking what made me think he was opposed to hierarchies, and pointing to one of his own papers in which he, too, had alluded to biological hierarchies.

I mention our connection with Dawkins in some detail here, for two reasons. First, conversation may be heated, and at times even turn nasty, around the High Table. But the spirit of open discussion generally prevails. The more imaginative participants—and Dawkins is one—prefer to gain the upper hand by force of argument rather than suppression. The peer review system in science can work to maintain and even improve both standards and integrity. But it can be all too easily abused.

Secondly, that Dawkins, on one side of the Table, and we naturalists on the other, agree that biological nature is hierarchically arranged underscores the true nature of our differences. We naturalists see the existence of large-scale biological systems, and the interrelationships among them, as crucial to understanding the context of evolutionary processes. Dawkins, instead, remains bent on seeing the complexities of biological organization and evolutionary history as a function of competition—among genes, or simply among organisms—for reproductive success. Both sides agree that complexity exists. Our differences lie in our approach to understanding that complexity.

Genealogical and Ecological Systems: Tradition and Innovation at the High Table

We are down to some very basic issues. We naturalists have been asking very simple, yet profound questions about the very nature of biological systems. Yet, as we shall soon see, some of the most cherished assumptions about various kinds of biological systems turn out to have been wrong—to the detriment of our ability to analyze the nature of the evolutionary process clearly.

For a start, let's take species. Species remain central, even critical, to evolutionary debate. In a very real sense, the debate crystallizes around two positions. We naturalists, taking up the threads of argument developed by Dobzhansky and Mayr in the 1930s and 1940s, see that the very nature and existence of species is central to the entire context of adaptive stasis and change. Without taking species into account, there can be no understanding of how that purring motor of natural selection actually works to produce adaptive stasis and change. Ultra-Darwinians, in sharp contrast, tend to downplay the importance of species and speciation. Some key proponents, such as George Williams, even deny that there is anything necessarily special about species at all.

That papers abound citing the numbers and average durations (in millions of years) of species is in part a reflection of the reemergence of the importance of species wrought by the nascent naturalist tradition. It is also, of course, a reflection of the mounting concern biologists and citizenry share for the accelerating loss of species—biological diversity—in the modern world. Perhaps it is true that nothing can so persuade one of the existence, the *reality*, of an entity more than its imminent demise.

Beginning with the work of Dobzhansky and Mayr, evolutionary biologists have seen species fundamentally as genealogical entities. Species are reproductive communities. Paleontologists have added a temporal dimension to the scheme. Their species are the lineages created by reproductive communities as they persist, generation through generation, through prodigious spans of time.

This is a purely functional construct of species. Species arise and are maintained through the reproductive proclivities of sexually reproducing organisms. Organisms make more of themselves, and in so doing set up lineages of ancestry and descent. Asexual organisms set up clonal lines, with each offspring genetically identical to its parent and sibs (save when mutations arise). Sexual organisms exist in and perpetuate breeding communities; they set up lineages that are stewpots of genetic variation.

Sewall Wright, as we have seen, painted a picture of species composed of subpopulations: demes. Demes come and demes go, with new populations splitting off from old ones, locally becoming extinct, or fusing back with neighboring demes. Sexual reproduction keeps demes going, and the vagaries of demic history are the stuff of the internal dynamics of species.

Sexual reproduction—organisms making more of themselves—underlies the phenomenon of demes, and ultimately of species. Genes replicate, making identical copies of themselves. Richard Dawkins has claimed that there are two kinds of biological entities: replicators and vehicles. In his scheme, genes are replicators and organisms are vehicles (necessary housing for the genes). As we shall see, a slight twist on this scheme actually underscores the core connection between the twin hierarchical worlds of biological economics and genealogy. There are, as philosopher David Hull pointed out early in the 1980s, two kinds of biological entities of evolutionary interest: replicators (as in Dawkins's scheme) and interactors.

More about interactors in a moment. Right now, replicators hold the floor, as they stand in the purview of traditional evolutionary concern. Genes replicate. Asexual organisms, by virtue of their genes, replicate. Sexual organisms—the vast majority of life (most microorganisms have some form of sexual reproduction intervening with asexual moments)—cannot possibly "replicate." True sexual reproduction involves halving a mixed bag of one's own genetic complement, and combining it with one-half of someone else's genes. No question of replication here.

That's why I prefer the informal, looser term "more-making." Sexually reproducing organisms may not replicate, but they certainly "make more of themselves." Genes make more genes. That's replication. Organisms make more organisms. That's not replication, but it is reproduction, or, simply, "more-making."

And species make more of themselves. Species "speciate." Species fragment, and parts of species, from time to time, have sufficiently independent histories in isolation that they diverge and become "new" species. Species make more of themselves, not so much like sexually reproducing organisms, but more like asexual (especially single-celled) organisms, that reproduce by fissioning.

That's the core of the idea: more-making. Genes may replicate, but the essential point for this purpose is that they simply make more of themselves. When they do that, in conjunction with cell division, they keep a multicellular organism's body alive. When they do that and make sex cells through the halving process of meiosis, which fuse to form a fertilized egg, they are supplying the genetic information for the production of new organisms.

Replication has shades of fidelity. Only asexual organisms replicate. In general, everything from genes on up through species "make more of themselves." And this is the internal dynamic that runs the genealogical hierarchy.

Genes make more of themselves; organisms make more of themselves; demes make more of themselves; and species make more of themselves. It is this more-making that creates and maintains the system at the next higher level. Organisms making more organisms keeps demes going; demal activity keeps species going. Species persistence and more-making—"speciation"—keeps large-scale taxa going.

Thus the "genealogical hierarchy": each level is composed of entities that keep on making more of themselves. If that action stops, if organisms stop reproducing, say, demes, and species stop existing. If species don't speciate, the continued existence of a larger scaled taxon (elephants, say: Family Elephantidae) depends solely on the continued survival of the extant individual species. The evolution of new species literally breathes new life into, and improves the survival probability of, higher taxa.

It is this constant activity, this constant "more-making" of the next-lower level components, that keeps entities at each rung of the genealogical hierarchy going. And all is in flux: the genome is subject to mutation and recombination as genetic information is handed down from generation to generation. Natural selection and genetic drift within demes and species further shuffles and rearranges the genetic deck—yet species remain generally unperturbed.

Ultra-Darwinians, then, are confused about (at least) two different aspects of hierarchies. George Williams, noting the chaotic variation and origination of adaptive diversity, fails to distinguish among levels. He fails to see, for example, that within and among local populations, a great deal of transitory evolution can occur without seriously modifying the overall adaptive diversity within an entire species. Williams therefore sees a conflict between the constant generation of adaptive change and stasis. He sees nature as a boiling cauldron seething with evolutionary change. And so stasis stands as a contradiction—and it does not help his grasp of the matter that he denies the very existence of species, at least as a "special" kind of biological entity. Were he to see species as entities composed of demes, themselves composed of organisms, he might conceivably have an easier time understanding that stasis and the constant change he is most concerned about are not antithetical. Great evolutionary flux within populations within a species is thoroughly compatible with the overall stability of that species taken as a whole.

But ultra-Darwinians at the High Table do not all share Williams's view that genes, organisms, and populations are essentially all that matter in evolutionary discourse. Richard Dawkins, as we have just seen, at least concedes the hierarchical structure of biological entities. Dawkins has been rather silent on the subject of species. He has talked at length of a very special category of higher-level entity: social systems (such as beehives and elephant herds)—a topic of such intense concern around the High Table that it warrants its own separate discussion in the next chapter. But, as an outgrowth of his concerns with social systems, Dawkins has also treated, albeit fleetingly, ecological systems. What little he has had to say of ecosystems is very revealing. It is here that we find our greatest differences.

The difference is simply put: Dawkins sees higher-level biological systems forming as an outgrowth of the competitive struggle for reproductive success. In his boldest prose, the struggle is between his selfish genes. Otherwise he is content to leave it as a more traditional Darwinian struggle for reproductive success among organisms. In his *The Selfish Gene*, Dawkins writes (p. 90):

> Maynard Smith's concept of the ESS ("evolutionary stable strategy") will enable us, for the first time, to see clearly how a collection of independent selfish entities can come to resemble a single, organized whole. I think this will be true not only of social organization within species, but also of "ecosystems" and "communities" consisting of many species. In the long term, I expect the ESS concept to revolutionize the science of ecology.

Maynard Smith's concept of "evolutionary stable strategies" was developed to counter group-selectionist arguments—cases in which selection was said to be acting for the good of an entire group, and not just its individual, component organisms. An ESS is an adaptive behavior that, as Dawkins himself defines it, if adopted by the majority of a population, cannot be bettered, improved upon, or displaced by some alternative strategy. It is evolved and maintained by individual, not group, selection. It is a useful construct, and is normally applied to situations of conflict within populations.

At issue here is Dawkins's claim that, in essence, the structure of communities and ecosystems arises as a simple function of individual-level natural selection. There can be no question that, within ecosystems, predators are adapted for capturing particular prey items, and their prey are adapted for avoiding particular forms of predation. African lions in the Namibian Kalahari Desert have developed a unique strategy for killing gemsbok (large antelope with long, pointed horns that make formidable defensive weaponry). Unlike the lions living anywhere else, Namibian lions will jump from behind, onto the back of a gemsbok to avoid the dangerous horns.

There is design in nature, and it is natural selection that effects the close fit between organism and environment. That is a credo not in dispute at the High Table, as I have made clear throughout. Ultra-Darwinians, as we saw in chapter 2, have transformed the older, passive construct of natural selection—as the winnower and keeper of relatively more successful variants—into a more active, functional vision. Natural selection has been equated with competition for reproductive success. Something is designing and holding together ecological systems—meaning two or more local populations of different species. Ultra-Darwinians such as Dawkins admit the reality of such systems. They also see design in them and would explain their existence through a relatively simple extension of natural selection theory. To an ultra-Darwinian, larger-scale biological systems are a straightforward result of that supposedly ever-present competitive struggle for reproductive success.

But there is a real problem here. An ultra-Darwinian may wish to see ecosystems as the products, the natural fallout, of competition for reproductive success. But an ecologist—someone who actually works on real ecosystems—sees them in entirely different terms. Ecosystems are interactive systems composed of local populations of (as a rule) many different species. It is the network of energy flow—between populations, and between organisms and the physical environment—that holds local ecosystems together. That, in any case, is the way ecosystem ecologists tend to approach their subject matter. It is a far cry from the ultra-Darwinian perspective.

As Stan Salthe and I pointed out in 1984, organisms do basically two (and only two) kinds of things: they engage in matter-energy transfer processes, and they reproduce. We called the former "economic" and the latter "genealogic." Reproductive behavior, as we have just seen, leads directly to the formation of larger-scale entities: the populations, species, and higher taxa of the genealogical hierarchy. Note, too, that it is not due to competitive reproductive behavior, but rather just the very fact of (sexual) reproduction itself that we have demes and species. And simple fragmentation of reproductive communities leads to the production of more species.

There are similar consequences of economic behavior. Members of local populations within a species variably cooperate, compete, or have neutral effects on one another as they lead their economic lives. That local population has a pronounced economic effect on the local environment—the local populations of other species, and the physical environment as well. The economic effect each such local population has, the role that it plays in the interlocking network of energy flow through the system, is what ecologists mean by the term "niche."

Local populations play definitive, and specifiable, economic roles in local ecosystems. The moment-to-moment economic interactions among organisms, whether with fellow members of their own species or with organisms from other species, has nothing directly to do with the reproductive concerns that sexually mature individuals have during the breeding season. We refer to local breeding populations as "demes." Paleobiologist John Damuth has suggested that another term—"avatar"—be used for local economic populations of a species. The two, of course, could be one and the same aggregate of organisms. A local population of a coral species, rooted as it is to the substrate, is both a breeding and an economic entity. Other local populations, however, change their compositions according to the time of year: sexually mature male African elephants are loners in all but the reproductive time of the year. Matriarchal hierarchies of mothers and young elephants form the herds that roam the countryside for most of the year. Avatars and demes are conceptually different entities, and their composition is often distinct as well.

As we have seen, reproduction—"more-making"—is the glue that binds components together to form larger-scale entities in the genealogical hierarchy. There is an analogue on the economic side: moment-by-moment economic interaction binds local populations (avatars) together, and it is energy flow between avatars that lends cohesiveness to these multispecies, cross-genealogical, entities we call "ecosystems."

Energy flows across local ecosystem boundaries, as well; thus, local ecosystems are joined into regional economic systems. Ultimately, all local ecosystems are joined into the biosphere—

the single, albeit complexly diversified, ecosystem of the entire earth that some have called "Gaia."

We are looking, simultaneously, at two rather different worlds. Every single organism is at once a part of two radically different, and hierarchically arranged, systems. Reproductively, an organism belongs to a deme, which is part of a species, which is part of an ancestral descendant skein of species that we call "higher taxa."

At the same time, an organism is leading an economic existence, is a part of a local population—an avatar—that is itself a part of a local ecosystem. That local ecosystem interacts with others on a regional basis to form still larger biotic, economic systems. By sheer dint of reproductive activity, large-scale genealogical systems are formed. By the same token, by sheer dint of economic activity, large-scale economic systems are formed.

This simple description of the organization of biological nature differs dramatically from the ultra-Darwinian view. Dawkins stresses competitive reproductive behavior among local conspecifics (that is, members of the same species) as the basis of social systems and even cross-genealogical systems. We hierarchy-minded naturalists, in sharp contrast, see demes and species as outgrowths of simple sexual reproductive behavior. Competition for reproductive success has nothing to do with it. We see ecosystems—composed of parts of many different species—as the simple outgrowth of economic behavior per se: whether competitive, antagonistic, mutualistic, or whatever. Economic systems exist simply because organisms lead economic lives, just as genealogical systems exist simply because organisms reproduce.

Organisms have behaviors, physiologies, and anatomical features that perform various aspects of their economic and reproductive functions. These are molded by the evolutionary process. They are, historically and evolutionarily speaking, adaptations. From an evolutionary standpoint, there is clearly a connection between the economic and reproductive worlds of organisms. But to spell out the nature of that connection, we first must accept the separateness of the economic and reproductive aspects of life. All organisms lead economic lives. Most organisms reproduce, and all organisms, of course, owe their existence to parental reproductive behavior. But the converse is not true: reproduction cannot go on in the absence of successful economics. When stressed, most organisms quickly stop reproducing, conserving energy and waiting for better times.

If Dawkins stops short of imputing genealogical properties to economic systems, he nonetheless does maintain that large-scale entities like ecosystems are the product of a process of competition for reproductive success. A description of a blatantly economic system in such reproductive terms means little to an ecologist. The connections between ecology and evolution must be sought elsewhere.

Evolutionists have even more commonly made the opposite sort of mistake as they have sought connections between the ecological and evolutionary realms. They have tended to impute economic characteristics to genealogical systems. Theodosius Dobzhansky seemed indignant when he sought to rebut the charge that evolutionary biology paid no heed to the fledgling science of ecology fifty years ago. And to some extent, Dobzhansky was justified: he, after all, spent long periods in the field observing and collecting fruit flies in the wild. He insisted that all his experimental observations be checked as far as possible among wild populations living in nature.

The formal connection that Dobzhansky and other evolutionary biologists since have seen between the ecological and evolutionary realms has not been so successful. Once again, species provide the prime example. It is commonplace to list the number of "species" present in a habitat, community, or ecosystem. Yet species, as entire entities, cannot possibly be present in such local settings, unless that species is restricted to but a single local population. We have no convenient shorthand to say "local representatives of a species" when we refer to the different sorts of organisms present in a localized environmental setting. John Damuth's term "avatar" is such a term, but hardly a household word, and not even widely known in biological circles.

Yet the problem is not simply terminological. We have already encountered George Williams's insistence that species are not "special" kinds of biological entities. The ingrained ten-

dency to regard species and populations as interchangeable per-
meates ecological thinking, as when evolutionists as well as ecologists see no problem with counting "species" present in an ecosystem. It also pervades the closely allied field of conservation
biology. Too often, "species" are listed as endangered when what
is endangered are local populations—parts of species. Sometimes there is a sociopolitical consequence of such errors: the
peregrine falcon, *Falco peregrinus*, was one of several raptor species in the United States suffering great reduction in numbers
from the use of DDT as a pesticide. Because it was listed as
endangered, effective lobbying and public outcry resulted in a
ban on DDT, and peregrine populations have begun to recover.
But in not revealing that peregrines have a near-worldwide distribution, conservationists expose themselves to the charge of
crying wolf. We must avoid saying that a "species" is endangered if we mean that a part of a species is under threat in one
particular place.

Species cannot be members of local ecosystems simply
because species are typically composed of many semi-independent populations. These local populations are loosely connected,
and they are integrated into different local ecosystems. The
North American coyote, *Canis latrans*, is a top predator in the
deserts of the southwestern United States, where it preys on
rodents, jackrabbits, and other small members of the southwestern fauna. In the northeastern United States, coyotes are forest
dwellers, eating small prey (they also take down an occasional
white-tailed deer), but their prey belong to completely different
species than those that their counterparts (other avatars) encounter in southwestern deserts.

In other words, maps of species distribution do not coincide
with the mapped distributions of ecosystems. Seldom do any
two species share identical geographic ranges. Yet we have this
concept that ecosystems are composed of parts, those parts being
portions of each of a number of different species. Avatars—not
species—are the parts of ecosystems.

Avatars, then, are the economic role-playing segments of
species. Avatars have niches. Species cannot have niches because

species do not play concerted roles, as entire units, in any kind of a natural economic system that we can specify.

So goes our naturalist argument. Our position challenges some rather cherished evolutionary suppositions about the connection between the evolutionary and ecological realms. That species have niches, for example, is no mere slip of the evolutionary tongue. As recently as 1982, Ernst Mayr, in the first major revision of his biological species concept since 1942, explicitly added the proviso that species have niches. I find this a distinct disappointment. Species do not play economic roles in nature. They do not have niches. Indeed, as Michael Ghiselin has summarized my argument, species do not seem to do much of anything at all.

Mayr's addition of his "ecological niche" addendum to an otherwise perfectly serviceable species definition was not the total aberration that it might seem to be at first glance. After all, niches have been attributed to species in informal terms pretty much since the term "niche" first entered the biological lexicon. But species as ecological role-players is just a part of the story of how evolutionists traditionally see the relation between ecology and evolution.

A larger game has long been afoot. Not only species, but monophyletic taxa—historical entities such as Mammalia and Trilobita—have long been alleged to occupy large-scale equivalents of ecological niches. Thus was the dilemma resolved. The ecological and evolutionary realms are connected because they are not really separate: genealogical entities—species and higher taxa—have ecological properties.

So thought George Gaylord Simpson and Theodosius Dobzhansky. Building on Sewall Wright's imagery of the "adaptive landscape," Simpson and Dobzhansky framed similar vistas of the history of life. Their vision yielded a seamless junction not only between evolution and ecology, but between small-scale microevolution and the large-scale macroevolutionary events of evolutionary history.

In the early 1930s, Wright proposed the metaphor of the adaptive landscape—a topographic map of sorts in which the beneficial gene combinations were represented as high points and less harmonious combinations as low points. The problem in evolution, as Wright posed it back then, was to maximize the number of individuals occupying the higher peaks, all within a species.

But Wright, in his original discussion, also alluded (however briefly and vaguely) to an entire species occupying peaks. As we saw in chapter 2, this was the sense in which Lewontin used the imagery in his *Scientific American* article on adaptation. As we have also already seen, George Simpson's concept of quantum evolution hinged on the problem of how a species could get from one peak to another in the adaptive landscape. Should those adaptive peaks be appreciably distant from one another (meaning quite different, as in his discussion of the shift from leaf browsing to grass cropping in early horse evolution), large-scale evolutionary change might conceivably be the result.

Elsewhere in his *Tempo and Mode in Evolution*, however, Simpson refers to the "adaptive zones" of large-scale taxa. They are the exact, larger-scale equivalents of the niches of species. But it was Dobzhansky who put it most eloquently, with a brief, pithy, and beguiling two paragraphs cementing the identity of ecology and evolution, and the smooth continuity of micro- and macroevolution:

> The enormous diversity of organisms may be envisaged as correlated with the immense variety of environments and of ecological niches which exist on earth. But the variety of ecological niches is not only immense, it is also discontinuous. One species of insect may feed on, for example, oak leaves, and another species on pine needles; an insect that would require food intermediate between oak and pine would probably starve to death. Hence, the living world is not a formless mass of randomly combining genes and traits, but a great array of families of related gene combinations, which are clustered on a large but finite number of adaptive peaks. Each living species may be thought of as occupying one of the available peaks in the field of gene combinations. The adaptive valleys are deserted and empty.
>
> Furthermore, the adaptive peaks and valleys are not interspersed at random. Adjacent adaptive peaks are arranged in groups, which may be likened to mountain ranges in which the separate pinnacles are divided by relatively shallow notches. Thus, the ecological niche occu-

pied by the species "lion" is relatively much closer to those occupied by tiger, puma, and leopard than to those occupied by wolf, coyote, and jackal. The feline adaptive peaks form a group different from the group of canine "peaks." But the feline, canine, ursine, musteline, and certain other groups of peaks form together the adaptive "range" of carnivores, which is separated by deep adaptive valleys from the "ranges" of rodents, bats, ungulates, primates, and others. In turn, these "ranges" are again members of the adaptive system of mammals, which are ecologically and biologically segregated, as a group, from the adaptive systems of birds, reptiles, etc. The hierarchic nature of the biological classification reflects the objectively ascertainable discontinuity of adaptive niches, in other words the discontinuity of ways and means by which organisms that inhabit the world derive their livelihood from the environment. (Dobzhansky, *Genetics and the Origin of Species*, 1951 (3rd edition), pp. 9–10).

It is a beautifully written passage, almost hypnotic in its allure. It seems so *right*. But, in the end, it makes little sense. If species don't occupy niches, then certainly higher taxa, which are strings of ancestral descendant species, surely cannot play concerted roles in some mega-ecological system. Dobzhansky's imagery confounds shared aspects of adaptive behavior among species within higher taxa with an imagined mega-economic role played by higher taxa. His choice of feline (cat) species was no accident, as individuals of many cat species share an over-arching stereotypic adaptive conformation of behavior and anatomy, particularly with regard to hunting. But that is all the pattern really is: a sharing among an assortment of species of organismic behavioral and anatomical traits. It is no different in kind than any other hierarchically distributed system of traits among species of a monophyletic lineage. And that is a far cry from some imagined concerted economic role that a taxon—a species, or any higher taxon—might be supposed to be playing.

There are connections between the ecological and genealogical realms—the dual systems that all organisms are willy-nilly caught up in through the mere circumstances of their existence. But evolutionary biologists have been slow to specify what these connections might be. We cannot simply invest traditional entities of evolutionary discourse—genes, demes, species, and monophyletic taxa—with ecological (economic) properties and think we have solved the problem. They have no such properties.

Dawkins, too, misses it when he asserts that ecological systems are by-products of competition for reproductive success, with fallout from the competitive reproductive drama within one species affecting the course of the drama within other species within the ecological system. Ecological systems function and are held together through moment-by-moment matter-energy transfer processes among avatars. A true physics of ecological systems examines the energy and information flowing directly through the system. Standing natural selection on its head, claiming that direct competition for reproductive success leads to competition for resources, which in turn governs all interactions between avatars within ecosystems is needlessly Byzantine. Organisms seek resources simply because organisms require such resources. Why they do the things they do, and why they seek particular resources with their particular bag of adaptive tricks, that's the evolutionary question. It is a quintessentially historical question.

Once again, we must look elsewhere for connections between the ecological and evolutionary domains.

Where the Twain Meet: Organisms, Populations, and Natural Selection

We have two separate worlds, separate realms of interactors and more-makers. On the one hand, we have an economic hierarchy of entities constantly engaged in matter-energy transfer process. On the other, we have an evolutionary hierarchy of entities whose components make more of themselves. They seem distinctly different, despite several attempts by evolutionists to see how they might be connected.

Yet they must be connected, and they are, in a deceptively simple way. Once again we return to organisms, those nearly forgotten entities in a world of biological research obsessed not only with genes and the other chemical constituents of cells, but also with species, monophyletic taxa, and ecosystems. Organisms loom as critically important only to developmental biologists, who long to learn the secrets of development—that interplay between inherited, gene-encoded information and the internal and external environment that sees a single-celled fertilized egg transformed into a complete, functional living being.

Organisms, we should never forget, are the only kinds of biological entity that truly are both more-makers and interactors. Only organisms actively seek energy and materials in order to be alive: to differentiate, grow, and maintain the living corpus. Only organisms reproduce. True enough, an organism's somatic (body) cells engage in matter-energy transfer functions; they also divide, producing more of themselves. But the cells of a multicellular animal, plant, or fungus are specialized, performing only a subset of all the physiological functions required for an organism to live. Only when organisms consist of but a single cell can we say that the hierarchy collapses and the cell/organism is both interactor and reproducer.

The naturalist perspective, as we have already seen, argues that the dual economic and genealogic (or ecologic and evolutionary) hierarchies of large-scale biological systems falls directly out of dual economic and reproductive pursuits of organisms. The division between economic and reproductive functions is even apparent within multicellular organisms. Towards the end of the nineteenth century, as biologists were trying to close in on an explanation of the principles of heredity, August Weismann produced his famous dictum. There is, he proclaimed, a disjunction between the cells of the body (soma—somatic cells) and those of the germ line. The relation between the two sets of cells is not symmetrical. Parental germ-line cells give rise to all the cells of the descendant organism. Any changes (mutations) in the genetic information of the germ-line cells stands to be inherited by the offspring. What happens, genetically speaking, to germ-line cells affects the organisms of the next generation. The converse, however, is not true: somatic cells have no impact on descendants. Mutations in somatic cells affect only an organism's own body.

Weismann's dictum is a one-way street, a unidirectional vector. The germ line affects the soma (of the next generation), but the soma does not affect the germ line. Weismann's insight laid to rest all claims of "inheritance of acquired characters" as an alternative evolutionary mode (though some possible, limited exceptions have recently begun to surface). Weismann's dictum has successfully been translated down a notch (in good reductive fashion) to embrace the far more recent findings of molecular genetics. The translation of DNA to various forms of RNA, and ultimately to the structural composition of proteins, is (in the vast majority of cases—we now know of exceptions) also a one-way street. The information in DNA is transcribed to "make" proteins, just as the germ line is responsible for generation of the soma. But the reverse seldom happens: information from proteins is not transcribed backwards (via RNA or not) to fashion sequences of DNA. This is molecular biology's "Central Dogma." The exception is "reverse transcription," which is known in some viruses ("retroviruses"), such as HIV (the viral agent that causes AIDS).

Weismann's dictum, if anything, seems to deepen the split between the economic and genealogical realms. It also proves to be half the key to understanding the two-way street that actually does connect the two realms. On the one hand, we have the obvious contribution that each generation makes to the ecological arena: organisms reproduce, and by making more organisms they are constantly producing new players for the ecological arena.

Think of what happens after an oil spill destroys a section of the intertidal zone along a coastline, or after a fire roars through a woodland, extinguishing all living things. Imagine, too, the colonization of wholly new areas, as when the volcanic island Surtsey appeared abruptly, the result of a spurt of volcanism along the mid-Atlantic ridge in 1963. In each case, ecosystems become established, often displaying classic sequences that ecologists call "succession." The point here is that the new system is established from preexisting demes/avatars, from populations living elsewhere.

That is the "function" of the elements of the genealogical hierarchy. Species are de facto genetic reservoirs. Ongoing generation-by-generation reproduction within demes constantly renews the supply of players—organisms—in the local ecological arena. Demes supply players to other areas undergoing colonization. Moose (*Alces alces*) were extirpated in the Adirondack mountains of upstate New York in the 1880s. They are now becoming reestablished, and the recruits are coming from several different regions in nearby Vermont and Canada.

But what of the opposite direction on this two-way street of connection between the ecological and genealogical realms? Is there no countermanding Weismann and the Central Dogma? Certainly there is: it is none other than natural selection itself. The connection is most immediately apparent if we paraphrase Darwin's original conception of natural selection: natural selection is the biasing effect that differential economic success has on an organism's reproductive success. We have already seen how the ultra-Darwinian preference for categorizing natural selection as simply "differential fitness" obscures the very existence of the separate economic and genealogical realms, and the important distinctions to be drawn between them. The ultra-Darwinian approach also masks the fact that the vector running from the economic to the genealogic side of the ledger—the only such vector—coincides precisely with the original Darwinian formulation of natural selection.

Dawkins, articulating the general ultra-Darwinian perspective, sees competition for reproductive success as the driving force behind the organization of ecological and social systems. He has it exactly backwards. Dawkins has the vector running the wrong way. It is the fate that organisms, with their heritable features—their economic adaptations—face in the economic arena that acts as the filter determining what proportions of genetic information are passed along to the next succeeding generation. Natural selection is that filter.

The point is critical. Evolution is a historical process, a record of the changing state of living systems. The genealogical systems—germ-line genes in organisms, demes, species, and monophyletic taxa—are organized entities of genetic information. Stasis and change—the fate of that information over time—is what is properly called "evolution."

Most of that genetic information pertains to economic matters: bodily features, including anatomies, physiologies, and associated behaviors that are the very economic adaptations of organisms. The genealogic side constantly reincarnates that information, each generation producing the organisms that play at the economic game. Because populations are limited, and because there is heritable variation in those populations, how well an organism fares economically will tend to have repercussions on its reproductive success. The vector is reversed, with the economic side "informing" the genealogic side what worked best in the immediate generation.

This naturalistic, hierarchical description of the basic structure and workings of biological systems is very different from the ultra-Darwinian perspective. They claim that competition for reproductive success drives competition for resources, which further structures and determines the inner workings of economic and social systems. We say that all organisms behave economically as a simple consequence of being alive; that the structural organization and inner workings of economic systems flows directly from such behavior; that economic systems depend on genealogical systems (reproductive behavior) purely as a constant supply of players in the ecological arena; and that what happens in the ecological arena helps determine the fate of genetic information as it is passed along from generation to generation in the genealogical context.

To Dawkins, germ-line genes—especially competition among them to leave more copies to the next generation—drive all aspects of life. Other ultra-Darwinians are somewhat less extreme, but nonetheless see competition for reproductive success as driving all biological systems. We naturalists, imbued with hierarchy, see neither system as driving the other. The two are largely separated, but nonetheless connected in reciprocal fashion. Neither is more important than the other. Evolution is history, not the gene-centered motor of and for the imagined constant change so dear to the ultra-Darwinian heart.

Why this ultra-Darwinian insistence that the biotic world is actually driven by competition for reproductive success? Organisms do indeed (sometimes) compete for reproductive success, and as Darwin pointed out in 1871, differential success falling out of such competition is distinct from natural selection. He called it "sexual selection." True natural selection, on the other hand, arises from competition for resources: economic competition. Why the ultra-Darwinian insistence that economic competition is really the same as, or a manifestation of, reproductive competition? For that, as we saw in chapter 2, is what the concept of "fitness" really implies.

But the mathematical convenience of lumping all sources of bias to reproductive success into a single catch-all rubric of "fitness" is not in itself sufficient to explain why the ultra-Darwinians have changed the sign of the vector that naturalists see as coming *from* the economic *to* the genealogical side. Something else is at work here, and that must be the ultra-Darwinian dissatisfaction with seeing evolution purely as historical ledger keeping. Science deals with interactions among entities—and all the interaction, except pure reproduction, seems to be going on in the economic sphere. Ultra-Darwinian desire has been great to deny the separateness of, the distinctions between, the economic and genealogical arenas. It is a desire to see evolution as an active process, like the sorts of dynamic processes studied by physicists, chemists, and ecologists. Evolutionists have consistently denied this separateness, have attributed economic properties to genealogic entities, and, most recently, as ultra-Darwinians, have insisted that competition for reproductive success drives economic systems.

What is this but physics envy? Claiming that economic systems are really all about the transmission of genetic information is a fairly blatant distortion of natural systems. More subtle problems arise when it comes to the biology—structural organization, function, and evolution—of social systems. That's where the battle lines are clearly drawn and crunch time comes in the High Table dialogue between the ultra-Darwinians and naturalists.

7

Paradoxes in Ultra-Darwinism

Sex, Social Systems, and the Reproductive Imperative

Over a half-century ago, a great debate raged around another High Table. Anthropologists were deeply divided over the relative virtues of an evolutionary versus a more analytic approach to understanding the functional dynamics of human sociocultural systems. The "functionalist" school, led by British social anthropologist A. R. Radcliffe-Brown, insisted that all one need do to gain complete understanding of a social system is to study all its components—essentially a description and analysis of the inner workings of the system itself. The past is sublimely irrelevant to the task of analyzing what a system is—what it is made of and how it works. In contrast, the evolutionists, with University of Pittsburgh anthropologist Leslie White as perhaps their most prominent spokesperson, argued that one can't understand the whys and wherefores of a culture unless one understands its history.

As far as I am aware, this old anthropological debate was never truly resolved. (I agree with George Williams when he says that scientific problems are not so much solved as quietly dropped in favor of some new set of issues that comes along to preoccupy a discipline.) But something very much like the old White–Radcliffe-Brown debate has surfaced recently in biological circles, particularly in the claims made by some developmental biologists led by Brian Goodwin of The University of Toronto. If the disputes I have been describing in this book take place around a metaphorical evolutionary High Table, the sort of disagreements between Goodwin and evolutionists amount to shouting between High Tables.

Goodwin frankly seeks a "physics" of developmental biology. His is the commendable goal of understanding how a fertilized egg turns into a fully formed adult organism—an understanding written in terms of the very physical and chemical processes that take place during that transformation. Why do vertebrates have, as a general condition, five digits at the distal end of the forelimb (that is, the "hand")? Evolutionists, Goodwin says, answer that the number five characterized the earliest tetrapods back in the Upper Devonian. It was the number that evolution "settled on" in the dim recesses of vertebrate history.

To a biologist of Goodwin's stripe, such a statement is true enough, but conveys as little satisfaction, as little true developmental understanding, as the statement "because God made it that way." After all, Goodwin has said, we don't seek a history of the solar system to explain why the planets are positioned as they are, follow the paths that they do, and have the speed and spin each has. All we need do is consult the laws of gravity to understand the present-day structure and inner workings of the solar system. It is Goodwin's fondest desire to see the equivalent realization take root in developmental biology.

Readers of this book will not have been struck by an overweening spirit of compromise. Yet in this instance, the supposed dichotomy between evolutionists, with their insistence on some form of historical explanation, and functionalists, who, impatient with history, seek to understand a system in terms of its own intrinsic structures and dynamics, are merely stating preferences for one or the other of two noncompeting, noncontradictory forms of causal explanation.

Ernst Mayr understood this well in an influential essay written in 1961. Mayr pointed out that the warbler flying south from his backyard one brisk night in the fall was prompted to do so by fresh gussets of cool northern air. The time was right to begin the migration south, and the bird's behavioral physiology simply reacted to the signal. But the signal is there in the first place because insect-eating warblers cannot make a living in northern climes when vegetation and dependent insect systems close down for the winter. There is nothing to eat, so the birds are adapted to leading migratory lives. A full explanation of that warbler's behavior encompasses both what Mayr called the "proximate" causes—physiological and behavioral functions triggered by definite environmental signals, and "ultimate" causes, the adaptations set in place through evolutionary history.

Evolutionists, of course, must have some grasp of structure and function. That's what adaptations are, after all: structural/functional systems fashioned by natural selection. On the other hand, all systems have histories. Even the evanescent traces of subatomic particles in a physicist's bubble chamber are histories of a sort. The history of the solar system has a lot to say about why the smaller planets occupy the inner reaches, while the very different giants lay beyond them. One's mode of explanation suits one's own purpose, but a full description of any system should include both its present state and whatever might intelligently be said about its past.

Why this preamble on rival approaches to causality in natural systems? Evolutionists have tried various routes to reconcile the apparent conflicts, or simply to bridge the gap, between the intrinsically historical nature of their discipline and the frank appeal of the functionalist approach. Mayr (and George Gaylord Simpson as well) has gone so far as to suggest that there is something inherently different about historical versus functional science. This is a dangerous and misleading path, in my view, as it suggests that subject areas such as evolution and historical geology fall outside the domain of "real" science.

Simpson's own work shows that "historical science" is no different in principle from "functionalist" science. His use of his-

torical pattern to evaluate conflicting schemes of causal explanation, culminating in his own preferred theory of quantum evolution (early version!) is the clearest case in point. For just this reason, I can see no deep distinction between historical and
functional science.

As we saw in chapter 2, a common pursuit in evolutionary
biology is actually a form of functional biology in disguise. Evolutionists will often perform an analysis of the structural and
functional properties of a system and then turn around and conclude that they are actually studying adaptation. The structural-
functional properties of an anatomical system—the structure of a
woodpecker's head, for example—is analyzed to answer the
question: Why doesn't a woodpecker blow its brains out as it
jackhammers a tree (or even a tin roof) with rapid blows of tre-
mendous force? Not unreasonably, the cushioning devices
revealed by such analyses are taken as adaptations. The point
being that an awful lot of functional biology is done in the name
of "evolutionary" biology. Nor is there anything intrinsically
objectionable about this line of work, so long as the crasser pitfalls of "just so stories" are avoided in the process.

Throughout this book I have been accusing ultra-Darwinians of going a step farther, of essentially leaving the traditional
camp of historical explanation, abandoning it for the beckoning
reaches of functionalism. I hasten to add that the dual hierarchical scheme that I myself prefer is also heavily structural-func-
tional. In my scheme, the elements of the genealogical hierarchy
are set up and maintained through ongoing "more-making,"
while the elements of the economic hierarchy reflect the
moment-by-moment interactions of their components.

What, then, are the differences between naturalists like
myself and the ultra-Darwinians on the other side of the High
Table? Just this: I see evolution as the fallout, the stasis and
change in the genetic information ensconced in genealogical systems. Genealogical systems are passive reflectors of what
worked and what didn't in the economic arena. In this, I am
articulating in modern terms an essentially deeply traditional
view of evolution as history. Just history.

Ultra-Darwinians are doing something else. They have embraced a thorough-going functional approach by transforming the concept of natural selection from that of a passive reflection of "what worked better than what" to an active determinant of functional process. They have changed the sign of the vector of natural selection: biological systems are structured, driven, and powered through an ineluctable competition for reproductive success.

Such a description, particularly of economic systems (as in Dawkins's suggestion that ecosystems can be interpreted in terms of competition for reproductive success) is rather strikingly different from the description afforded by our naturalistic description of dual economic and genealogical hierarchical systems. It is rather easy to counter an argument that describes the structure and function of ecosystems in terms of reproductive competition.

More subtle is the realm of social systems. Ultra-Darwinians define social systems essentially as reproductive cooperatives. In contrast, to my naturalist eyes, social systems are neither purely economic nor purely genealogical systems—a position I have recently explored in depth with philosopher Marjorie Grene, in our book *Interactions* (1992). We see social systems as intriguing hybrids, with economic and reproductive elements complexly interwoven. Social systems evolve—have phylogenetic histories—like any other sort of biological system. But they can be described at length in purely economic and reproductive functional terms. I prefer to speak of the "biology" of social systems generally, embracing both functional and historical aspects.

Sociobiology—essentially an evolutionary biology of social systems—is the outstanding achievement of ultra-Darwinian biology. At the very least, it is the field in which the application of ultra-Darwinian principles has attracted the greatest attention. It is a subdiscipline unto itself. It has made great inroads even into anthropology. This is a reflection, I believe, of the general cultural infatuation with gene-centered explanations that has become one of the legacies of the molecular biological era. In any case, sociobiology is the centerpiece of ultra-Darwinism.

To a sociobiologist, a social system arises and is held together as an outcome of competition for reproductive success. All social behavior, including economic behavior, arises from or is seen as an adjunct to reproductive behavior.

Precisely because so much of social behavior is about reproduction, sociobiology has contributed a great deal to what I prefer to call the "biology" of social systems. Sociobiology has spurred an inordinate amount of field research—much of it very good, and nearly all of it badly needed survey work—on social systems. Performed in the name of evolution, nearly all this work is in reality purely functional biology—reproductive biology, to a great degree, but also economic (ecological) biology examined in the name of the reproductive imperative.

All sociobiology is done in the name of evolution. Very little of it is truly evolutionary in the sense defended here: the historical development of social systems. There are exceptions, of course. E. O. Wilson, as close to a founding father of the sociobiological movement as anyone (he gave it its name and its initial exegesis in *Sociobiology*), has written penetratingly on aspects of the evolution of insect societies.

But most sociobiology remains purely functional biology. The hierarchical perspective on social systems is likewise functional in spirit. Ironically, it is over quintessentially functional issues—descriptions of the components and inner workings of social systems—that High Table discussion is joined. The overall context may be evolutionary, but the disagreement at the High Table is really over the functional biology of social systems.

The Genesis and Structure of Sociobiology

Creative minds love a good paradox. Ultra-Darwinians have confronted two during the past 30 years. In one case, they feel, they have scored an unmitigated success. Resolution of the paradox of altruism led directly to the sociobiological explosion. What is the paradox of altruism? Natural selection hones adaptations that are "for the good of" organisms, and not for the good of species. How, then, to explain altruism where individual organisms seem to be acting in the interest of others, sometimes even at their own expense? Altruism is especially apparent in social systems, which seem to be structured around a high degree of cooperative behavior among members.

The second paradox is the very existence of sex. If the very name of the game of life is to leave as many copies of your genes behind, why on earth do so many organisms reproduce sexually, thereby limiting themselves to contributing only half of the genes that go into each descendant? Here, ultra-Darwinians rather candidly admit, they haven't quite got it all figured out. Despite a number of books (including one each by George Williams and John Maynard Smith) and multiauthored symposia, the paradox of sex has not yet been resolved. Nor will it just go away.

But first things first; back to altruism. Oxford University biologist William Hamilton cut through the Gordian knot of altruism with a neat double stroke—two related papers published back-to-back in the *Journal of Theoretical Biology* (1964): "The Genetical Theory of Social Behavior," Parts 1 and 2. In a nutshell, Hamilton proposed that the manner in which organisms behave towards one another is a function of their degree of genetic relatedness. The more genes two organisms share, the more likely they will cooperate and exhibit altruistic behavior. Thus the notion of "kin selection": an organism's own fitness is enhanced if it contributes in some way to the reproductive success of a close relative. A sibling hanging around a nest helping to raise a sister's brood is indirectly helping see to it that the sib's own genetic makeup is to some lesser degree also being perpetuated. A cousin doing the same thing would reap rather diminished rewards: collateral kin such as cousins share far fewer genes than do sibs. Ronald Fisher is said to have figured much of this out on the back of an envelope in an English pub over half a century ago.

Thus, to an ultra-Darwinian, the apparently altruistic cooperation underlying social behavior is really a form of selfishness in disguise. Cooperation comes in degrees, depending utterly on the extent to which cooperating individuals share their genes.

It will come as no surprise, then, that the ultra-Darwinian approach to social systems works best in those systems where all the members are directly related. E. O. Wilson recognizes four "pinnacles" of social systems: marine invertebrate colonies, insect colonies, nonhuman vertebrate (Wilson actually says "mammalian") social systems, and human societies.

Colonial marine invertebrates present the purest situation. All the individuals (polyps) of a colonial coral or jellyfish, or the individual zooids of a bryozoan colony, are genetically identical (save for any mutations that may arise in individual polyps). Coral and jellyfish are members of the Phylum Coelenterata. All coelenterates are very simple animals. Individuals lack true organs—which serve to divide labor within the bodies of more complex animals. Bryozoans—tiny creatures with tentacles that sweep passing waters for nutrients—are somewhat more advanced.

Colonies of both corals and bryozoans are initially established through normal sexual reproduction. An egg is fertilized and develops into a single individual called, respectively, a polyp or a zooid. Subsequent development of the colony, however, is purely asexual: new polyps or zooids bud off preexisting ones. Most coral colonies remain fairly monotonous, with but one kind of polyp present throughout. Each polyp feeds, respires, and is capable of reproducing (both asexually and by shedding sperm and eggs into the surrounding sea water).

Colonial jellyfish and bryozoa are another story. Here the colonies might consist of many different kinds of polyps or zooids. The Portuguese man-o-war, for example, consists of special polyps modified into a flotation bubble and various feeding, reproductive, and defensive polyps, including the stinging polyps that immobilize prey and ward off predators. Bryozoans are similar. A colony may have several different types of zooids, including some that feed, others that bear the reproductive duties, and still others that clean the colony and ward off enemies.

Division of labor is the key to understanding the biology of social systems. A complex jellyfish or bryozoan colony is, in many fundamental ways, very much like a single multicellular animal organism. Think of it: the cells—billions of them arrayed into some 200 different cell types in a human body—are nonetheless genetically identical (save whatever mutations appear). We see individual polyps of some coral species living on the seafloor and conclude that colonies of corals are truly collectivities of many different individual polyps. But it is just as correct, and perhaps more apt, to see such colonies differently, as analogous to a multicellular animal body. The division of labor among different polyp types is exactly analogous to the diversified organ systems of a complex animal body.

We have no trouble seeing that the diversified cells, tissues, and organ systems of our own bodies work in concert for the "common good" of ourselves—the organism. They work together so that the organism survives, and so that the organism can reproduce. With marine invertebrate colonies, much the same situation prevails. There is no conflict of genetic interest: it is "all for one and one for all." The colony itself is an individual, and the division of labor ensures an economic and reproductive life for the colony as a whole.

Are the hives of social insects in any way similar to colonial marine invertebrates? Sociobiology sprang from insect biology as much as it did from theoretical considerations of the nature of altruism. E. O. Wilson, among many other things, is an ant systematist. Ants are hymenopterans, members of a huge insect order that also includes bees and wasps. Many species are social. Crucially, there is an entire spectrum ranging from totally asocial—essentially a solitary existence—through modest rather simple colonies, through highly complex colonial systems—with great diversification into a number of castes and with a great division of labor—through megacolonies—where many such complex colonies are welded into a large supersocial system.

Such spectra of variation are always prime evidence for evolution. Wilson, moreover, has shown that complex social systems have evolved within at least 11 separate lineages among living Hymenoptera.

Bees and wasps display a form of inheritance known as "haplo-diploidy." All members of a hymenopteran colony are related, but to varying degrees. A queen's unfertilized eggs develop into males. They are "haploid," with only one set of genes. Females, on the other hand, develop from fertilized eggs. With two sets of genes, they are "diploid." The odd consequence is that a female is more closely related to (meaning shares more genes with) her own sisters than to anyone else—her mother, father, brothers, or her own children.

Wilson has described how, in many hymenopteran social systems, an individual will change its role—its caste—as it progresses through life. The early days might be spent tending eggs or larvae, while later stages involve leaving the nest in search of food for the benefit of all stay-at-homes: soldiers, workers, queen, and larvae. A signal victory for Hamilton's notion of kin selection is that the application of economic behavior to aid in the reproductive task (as in the care and feeding of larvae) is observed to be more or less proportional to the degree to which the helper shares its genes with the larva.

Nonetheless, insect social systems are far more similar to marine invertebrate colonies than they are to vertebrate social systems. The reason is this: in marine invertebrate colonies, and in insect colonies, some individuals are doing the reproducing, while the others are diversified into a number of different economic tasks. All members of the colony are either genetically identical (invertebrate colonies) or are related to some varying degree (insect social systems). Though it is by no means "all for one and one for all" in a hymenopteran hive because of the asymmetry of the genetic relationship among all the members, nonetheless the form of economic and reproductive cooperation reflects the common genetic interests in very much the same fashion as found in colonial marine invertebrates. Vertebrate social systems, in contrast, only rarely show such stark division between economic and reproductive behavior.

Sociobiologists treat insect social systems as reproductive cooperatives; all economic behavior ultimately supports reproductive activities. This includes such obvious behaviors as care and feeding of larvae, but also the less obvious storehousing of foods for general colony consumption. To my naturalist eyes, it is as much a distortion of reality to say that an insect colony is organized around reproduction as it is to say that a complexly-organized jellyfish—or the human body—is organized purely around a supposed reproductive imperative.

There is another consideration as well—one that arises from a naturalistic, hierarchical perspective. As associations of many organisms of the same species, social systems invite comparison with both demes and avatars. (Demes, recall, are local breeding populations; avatars are local populations of a single species that play a particular role in a local ecosystem.) Viewed from the vantage point of their relation to the surrounding world—to local ecosystems and to other colonies of the same species living elsewhere—the answer to the question, Is a social colony a deme or an avatar? must be yes! A hymenopteran hive is *both* a deme and an avatar.

There is no question that a local colony of wasps exerts a concerted force, plays a concerted ecological role, in a local ecosystem. There is also no question that reproductive cooperation is a sine qua non of such systems: hives are localized demes. Thus social systems have a very special status. They are the only biological system above the level of the organism that can realistically be said to play both an economic and genealogical role in nature. Insofar as marine colonies and insect social systems are concerned, this hierarchical view only underscores their status as "megaindividuals." They are structured and act more like the complex bodies of individual multicellular animals than like a loose cooperative of many different individuals. One thing is certain: an insect or marine invertebrate colony is no more purely about reproduction than is anyone's individual body.

What, then, of vertebrate social systems? These are arranged in a vastly different manner from the division of labor in marine invertebrate and insect social systems. In an insect colony, a few individuals reproduce, while all others perform other, separate, economic tasks. Not so with vertebrates: males and females generally lead both economic and reproductive lives.

The African naked mole rat is virtually alone among vertebrates in mimicking hymenopteran social systems, with one dominant queen doing all the colonies' reproductive chores.

What is more, vertebrate social systems, in marked contrast to their invertebrate counterparts, usually do not consist exclusively of closely related individuals. Thus it is even more difficult to view a vertebrate social system purely as a reproductive cooperative. Vertebrate social systems are, on the face of it, as much economic as they are reproductive cooperatives. Kin selection looms as a much less important facet of vertebrate social existence than it does in invertebrate systems (insects, marine invertebrates) where there is a degree of genetic relatedness linking all members of the system.

Nonetheless, kin selection has been invoked to explain aspects of vertebrate social organization. Older offspring, among some hyenas, for example, have been found to help rear succeeding broods. More famous is the example of scrub jays living in (now endangered) scrub oak communities in Florida. Older offspring among Florida scrub jay families also hang around, helping in various ways to rear their parent's later broods. Originally cited as a clear-cut example of kin selection, the two biologists, G. E. Woolfenden and R. E. Kirkpatrick, who reported an exhaustive study of these birds, preferred to interpret the phenomenon in a more classically Darwinian fashion. Older scrub jay siblings, they concluded, hang around and help raise younger sibs, not so much because the younger sibs will help transmit the older sibs' genes vicariously to the next generation, but rather because hanging around and helping mother and father puts them in line to inherit the much-sought and hotly contested breeding territory. According to Woolfenden and Kirkpatrick, older sibs help their parents with later broods for directly selfish, albeit reproductive, reasons. They want mom and dad's chunk of real estate to set up their own reproductive shop.

Well and good. All social systems are indeed concerned with reproductive affairs, but they are not exclusively about reproduction. At stake with the Florida scrub jays seems to be mom and dad's territory. It is a breeding territory, to be sure; but it is first and foremost pure territory—where each scrub jay lives out the vast bulk of its economic existence. Scrub jays live and die in and around those territories. Indeed, female birds in general are now known to be judging prospective mates not so much on the glories of their mating calls or brilliant plumage as on how inviting the territory controlled by each male looks. That means how easy it is to live there: how available foodstuffs are, and how safe the territory really is. That is an economic matter.

It boils down to this: If you embrace the ultra-Darwinian perspective that all economic activity is performed for the express, ultimate purpose of passing along genes, the economic component of social systems emerges as subservient to the goal of maximizing those genes in the next generation. This goal might be achieved through helping close relatives reproduce, but passing genes along is the central goal, the raison d'être, of any organism's life.

If, however, you adopt the more balanced view—the naturalist perspective that economics and reproduction are the two fundamental, yet largely separate, aspects of an organism's life—you will inevitably see social systems in a rather different light. You will see social systems not as dynamic reproductive cooperatives, but rather as complex meldings of the economic and reproductive behaviors of organisms.

Human social systems offer an instructive case in point. Culture—learned behaviors transmitted through language, rather than genes—is virtually unique to our species, making the phenomenon of human sociality very different from all other forms of social systems. Traditionally, anthropology and sociology have received little input from biology.

All that has changed in the last 20 years. The biological invasion of social science turf began in 1975 with the publication E. O. Wilson's *Sociobiology*, with its notorious "Chapter 20" on human social behavior. Wilson did not single-handedly invent the recent rush to reduce virtually every facet of human existence to genetic terms. But he gave the movement strong impetus with his suggestion that all manner of human behaviors—from schizophrenia to religion to criminality—have a strong genetic component, and are thus to be construed in ultra-Darwinian terms.

This modern guise of the nature-nurture debate, however riveting, is really a side issue at the High Table. Few High Table-ites on either side of the evolutionary debate are extreme genetic reductionists when it comes, that is, to explaining human behavior and sociality. Ultra-Darwinians, as a general rule, avoid asserting that there is a gene for religious activity. On the other hand, few at the High Table (on either side of the issues) deny a genetic component—true heritability—underlying many aspects of human behavior.

More relevant is the simple question: how well does the ultra-Darwinian-inspired sociobiological paradigm describe human sociocultural organization? The answer: not very well at all. The reasons illuminate the more general failures of this body of thought. By picking the oddest, most extreme case—human sociocultural behavior and organization—much light is shed on the interactivity between economics and reproduction in social systems generally. The failure of sociobiology to provide an adequate account of human sociality reveals its general weakness, its distortion, in describing all other social systems.

We humans have, interestingly, partially decoupled our reproductive from our economic behavior. Recall that basically nonsocial animals lead largely separate economic and reproductive lives. In social animals, there is more direct interplay between economics and reproduction—as when the economic behavior of workers in an ant colony has direct reproductive consequences, such as in the tending of larvae in the communal nursery.

In humans, there is likewise much interplay between the economic and reproductive domains. Nuclear families definitely seem a universal in human experience, despite occasional counterexamples (Israeli kibbutzim) and recent changes in family structure in the United States. Nuclear families are complex economic and reproductive cooperatives. But there is a novel twist complicating the economic-reproductive dance in human sociocultural behavior: sex.

Humans are sexually reproducing mammals. To a degree unparalleled in any other known sexually reproducing species, sex and reproduction are significantly decoupled—meaning they are functionally separated in a very important way. It takes sex to make kids. But the reverse is not true: unlike virtually all other species, it does not take reproduction to have sex. Sexual behavior exists in and of itself, and often serves as the express link between economic and reproductive aspects of human life.

Consider the nuclear family. One scenario for the origin of this fundamental human sociocultural phenomenon sees permanent pair bonding among humans as basically a food-for-sex arrangement. Females evolve permanent sexual receptivity, keeping males around on a regular basis. Males, in turn, provide crucial economic input—food. The complex reproductive-economic cooperative that is a nuclear family develops. This, the barest bones of a scenario sketched by anthropologist Helen Fisher in *The Sex Contract* (1982), is of course a vast oversimplification. But it does capture the flavor, in explicitly evolutionary terms, of the nature of the relationship between economics, reproduction, and sex as a distinct third ingredient in human social behavior.

Most of the elements of human social organization in modern, complex societies are explicitly economic. Firms—paragons of hierarchical organization—are all about human economics. So, for that matter, are scout troops and all similar organizations that promote social skills. Sociobiologists can point to what they take as the ultimate meaning of such economic pursuits: wages and socialization are necessary adjuncts that merely contribute to the real, ultimate goal—reproduction. What is missed in such a perspective is that the systems stand on their own. The human experience is the best possible demonstration that, even within a social context, it is madness to interpret all forms of economic behavior strictly for the implications they may have for the reproductive success of the participants.

Reproduction is far from the immediate concerns, and often the immediate consequences, of such economic behavior. Sex, interestingly, is by no means so far removed from economics. The connection between economic power and sex is legendary. There are direct economic implications of sex—and sexual implications of economics, where reproduction is literally besides the point. Fascinating as the details of the relation between sex and economics in human affairs [sic] may be, it is also rather beside the point of High Table discussion. With one fundamental exception: homosexuality.

Homosexuality is a real paradox to ultra-Darwinians. Anyone who believes the whole point of existence of organisms is to struggle to pass on as many copies of their genes as possible is going to have trouble accounting for homosexuality. It is a complex issue. The particular form of the nature-nurture debate surrounding homosexuality lies in the either/or question: Is homosexuality a preference, a "choice," or is it biological manifest destiny, dictated by genetic inheritance? The political implications of the answer to this question are hotly debated, with some homosexuals preferring genetic determinism, others not. Crucial here are the implications of homosexuality purely from the perspective of evolutionary biology.

Ultra-Darwinians (including some homosexuals) have no ready explanation for homosexuality. Kin selectionist arguments have surfaced from time to time, and there are, of course, many examples of homosexuals aiding family members who themselves may eventually reproduce. But kin selectionism offers but faint hope of a full accounting for human homosexuality in ultra-Darwinian terms.

Homosexuality is the perfect illustration of the decoupling of sexual from reproductive matters so characteristic of human behavior. It also illustrates how sex and economics are linked. If kin selection is basically not relevant to explaining the prevalence of homosexuality, economic-sexual cooperatives definitely are. Whether it be transient relationships, or the emerging phenomenon of more stable relationships increasingly recognized by the power structure of society, homosexual relationships are economic-sexual in nature. They exhibit all the trappings of male-female pair bonding, sans reproduction. Indeed, the increasingly common phenomenon of adoption among homosexual couples only reinforces the general conclusion: homosexual pair bonds can have all the trappings of male-female nuclear pair bonds, including the rearing of children, without the need to spread one's own genes to the next generation as the driving factor. I do acknowledge recent reports of reproductive activity (as through artificial insemination) on the part of one (or even both) partners, and that homosexuals might also reproduce in earlier relationships, before entering homosexual relationships. These exceptions by no means contravene the more general situation: homosexual relationships are typical, "normal" human relationships in every respect. They just cut out the reproductive side of things. So too, as a coda to this discussion, are some heterosexual unions—fully economic, fully sexual, but, by mutual decision, without children. Madison Avenue even has an expression for it: DINKS—double income, no kids.

DINKS and homosexuals are the exceptions that literally prove (meaning "test") the rule. It is ludicrous to maintain that the social system that we know best—our own—is absolutely and fundamentally driven by the need of each and every one of us to leave as much of our genetic information as we can to the next generation. The human world, and the biological world in general, just doesn't work that way. It is more complex. It is not as "tightly wrapped" as the hyperdeterminism of ultra-Darwinism wishes it to be. Life is simply not solely about the transmission of genetic information. It is also about the economic affairs of actually being alive.

All of which takes us face to face with that deepest ultra-Darwinian mystery, the origin and maintenance of that manifestly inefficient reproductive mode: sex.

Why Sex?

When an animal body (somatic) cell divides to form two daughter cells, all the chromosomes, with all their component DNA sequences, are duplicated, lined up, and pulled apart, an identical set going to each cell. The process of supplying each descendant cell with an exact copy of the parental cell's genetic makeup is known as mitosis. Animal cells are *diploid*, meaning that the chromosomes come in pairs, which are not identical. The same basic genes occur on each identical pair, but often come in slightly variant forms—alleles. In mitosis, descendant cells receive a copy of both versions of each chromosome.

Not so when sex cells are formed—the process of meiosis. Meiosis starts out like mitosis, but there is an extra step: the two descendant cells divide once again, only this time the genetic material is cut in half. Each of the four sex cells (four sperm, or one egg and three "polar bodies") gets only one chromosome of each pair. Genetic information is pulled apart and recombined in a novel form in sexual reproduction. The result is one source of the variation in a population within each generation. Variation is the raw stuff of evolution, the very material on which natural selection acts.

Thus did sex seem a natural part of the evolutionary drama. Few biologists saw the existence of sex as a problem for the first 100 years or so after the publication of Darwin's *On the Origin of Species*. Sex presents variation, and variation is good for the survival of the species, or so it was thought. Dobzhansky certainly thought so, seeing a neat counterbalance between selection always trying to eliminate variation to perfect adaptive fit to the environment, and the countervailing production of variation so that a species would not become overly specialized and face extinction if the environment changed. Geneticists ever since Herbert Muller in the 1930s tended to see sex as a good thing, affording the variation that enabled rapid spurts of evolution.

But that was before the purism espoused by George Williams in his *Adaptation and Natural Selection* (1966) fully took hold. As we have seen, perhaps the most fundamental credo of ultra-Darwinism is that selection—meaning natural selection in the strict sense—can have no "eyes" for the future. Selection cannot be for the "good of the species." It can only be a measure of what works best for individual organisms in their struggle for existence. In that, of course, Williams was perfectly right. Natural selection is all about differential reproductive success among individuals within a local deme. It reflects what worked better than what in a local population of individual organisms. It is not about the survival of species, but about the survival of organisms and the consequences for their reproductive success.

All of a sudden sex became a big problem. That central claim of ultra-Darwinism—that nature is organized around a competitive compulsion to leave as many copies of one's own genes to the next generation—is incompatible with the mere existence of sex. If the name of the game is to leave as many copies of your genes behind as possible, it is pure folly to mix them with someone else's on a 50-50 basis. Some organisms, in fact, don't do that, reproducing asexually. They should be favored by evolution, but clearly they are not. Even bacteria, which lack true meiotic sex, nonetheless freely exchange bits of genetic information. Some few groups—the minute, enigmatic tardigrades ("water bears")—appear to be wholly asexual. And some sexually reproducing lineages occasionally produce asexual descendant species, such as some asexual clones ("species") of whiptailed lizards. None of these have inherited the earth, nor is it their destiny to do so.

Sex, ultra-Darwinians have reasoned, must be for something. But the old explanation, that sex supplies the variation necessary for persistence and further evolution of a species, was no longer available to them. Why, then, is there sex?

To their credit, ultra-Darwinians have not shied away from this conundrum. They argue—forcefully, yet good-humoredly—among themselves as they sit across from us at the High Table. I have already mentioned that John Maynard Smith, George Williams, Michael Ghiselin, and several others have written books wholly or mostly devoted to this topic. A recent volume, *The Evolution of Sex* (1988), edited by biologists Richard E. Michod and Bruce R. Levin, exposes the nature of the within–ultra-Darwinian debate about sex. Sex remains a great ultra-Darwinian preoccupation!

What are they saying? The details of ultra-Darwinian internal wranglings on the subject are labyrinthine, with the main point being that they cannot agree. Some, like Maynard Smith, think that there is something of a group selectionist argument to be made, that there need not be a short-term benefit to organisms to be found for the persistence of sex. Against this (as Maynard Smith admits) is Williams's argument that there must be a short-term benefit of sex or it would be suppressed by the evolution of parthenogenesis, where unfertilized eggs develop utilizing only maternal genetic information. Other biologists have suggested a variety of such short-term, direct benefits of sex. Most attractive (to me, that is) is the idea that diploidy—having two nonidentical copies of each chromosome—facilitates DNA repair. Mutations that might arise and prove harmful to the organism might be nipped in the bud, suppressed and corrected by internal cellular repair mechanisms enabled by the presence of an alternative gene. Sounds plausible to me (though it must be borne in mind that I am a paleontologist), but Williams says that he, Maynard Smith, and a number of other ultra-Darwinians find the argument too "extreme."

We naturalists have sat as largely silent witnesses to this internal ultra-Darwinian debate. We have occasionally interjected a few observations and suggestions—which, so far as I can tell, have fallen on mostly deaf ears. First to enter the fray was paleobiologist Steven M. Stanley, who suggested that sex leads to divergent speciation, thus keeping clades alive. Asexual clones diversify too slowly to overcome normal extinction rates. Stanley's discussion incorporates a sense of hierarchy, but otherwise is really an extension of the old Muller argument that sex enables evolution to occur—thus beating back the threat of extinction.

I prefer my own suggestion, offered only as an additional perspective and differing only slightly from Stanley's. It is one that is also fundamentally hierarchical. It is a purely functional postulate: sex prevails because sexual reproduction creates the existence of demes and species. And species are notoriously difficult to dislodge. Present as local avatars in (usually) a wide variety of local ecosystems, species are extinction resistant. It is not that sex promotes rapid evolution so much as that species resist extinction. The argument is superficially circular: sex implies the existence of species, and the existence of species implies the continued prevalence of sex. It is not so much circular, though, as it is a feedback loop linking discrete levels of the genealogical hierarchy.

Yet all of this discourse on the continued prevalence of sex may be, in a very real sense, beside the point. After all, sex is only a problem if we assume that organisms are locked, willy-nilly, into an eternal competitive struggle to leave more copies of their genetic information to the next generation. Once we admit that nature might not be so dogmatic about the reproductive imperative, sex ceases to loom as such an intractable problem.

Enough. Time to draw back, take stock, and sum up where the High Table debates between ultra-Darwinians and naturalists stand.

Epilogue

Whither the High Table?

Scientists are supposed to hold their theories lightly, to explore all opportunities for knocking them asunder, to "falsify" them. That, it is commonly said, is the way that science progresses: by showing that something is wrong. If an idea squares with the facts (what we think we know to be true), then it might be right. If not, then that beautiful idea is indeed slain by an ugly little fact. We are far more certain when an idea is wrong than when it is right.

That is the nature of things, but no single scientist ever behaves in this proscribed fashion of self-doubt and flagellation. Scientists, being as a rule more or less normal human beings, passionately stick up for their ideas, their pet theories. It's up to someone else to show you are wrong. The scientific enterprise is a true collectivity. Through all the din and shouting, eventually some progress is made in achieving a better—a more accurate, more realistic—picture of nature.

Once, in my younger, more naive (and perhaps more idealistic) days, I was describing elements of High Table discussion to a gathering of mathematicians, chemists, and physicists. I told them that it must seem strange to them that such persistent disagreement, without any real prospect of a definitive solution one

221

way or the other, could go on seemingly forever in evolutionary biology. After all, I said, in chemistry and physics everything is resolved as the truth is steadily revealed in an endless series of elegant experiments. Or something like that.

I was nearly laughed out of the house, and after a quickly dissipated spurt of embarrassment, I was delighted to drop the myth that there is something wrong, something imprecise, something somehow unscientific about the messy field of evolutionary biology. Physicists can't agree among themselves either!

The wrangles at the High Table are hierarchically arranged. The grand split between what I have been calling ultra-Darwinians and naturalists has its roots in the writings of a single individual: Charles Darwin. Certainly, ever since the advent of genetics nearly 100 years ago, that split has been there in one guise or another—right up to tonight's supper at the High Table. But, within the major factions there are, inevitably, lesser wrangles. Ultra-Darwinians argue among themselves about the nature, meaning, and importance of group selection. They fight over the meaning of sex. We naturalists, on the other hand, argue over the nature, existence of, and significance of species selection. Nor are we unanimous on the causes (even the realities) of mass extinction, or their relevance to subsequent episodes of evolutionary diversification.

Even if some issues are resolved to some degree of general satisfaction, there is no guarantee that they won't arise, phoenix-like, a generation or two down the line, to be reevaluated, fought over again with fresh passion and, no doubt, to a different conclusion—if a conclusion there happens to be.

So, who is right? That cannot be the real question. I do passionately believe that there is a physical reality, a material universe. I do think that it is the job of "science" to describe that universe as accurately as possible. There is, in other words, a correct solution, a right answer. On the road to those answers (which, it seems, we will never fully recognize when and if we get them), all we have is successive approximations—competing claims to truth, always based on different "takes" on the nature of things. If science is always trying to obtain clearer snapshots of that physical reality, scientists are more like painters (or, for that matter, art photographers) whose tastes and objectives are bound to differ. No two painters render the same subject in precisely the same way. And so it inevitably is in science.

The history of evolutionary biology, it seems to me, is a history of successive waves of arriviste painters who claim that they, finally, have the truth. This succession of truth-knowers are most clearly tied (but not restricted) to technological innovation in biology, itself a litany of conversion of biology from the supposed imprecisions of old-fashioned natural history progressively approaching the true Nirvana—the physics model of a real science. In any case, discovery of Mendelian genetics was supposed to have been the death knell of Darwinism. But population genetics eventually sprang up and, eventually, effected a genuine rapprochement between the new "physiological" genetics and the older Darwinian vision. The discovery of the principles of genetics did not, it turns out, make the real phenomena that Darwin so ably described disappear. It seems absurd, in retrospect, that anyone would think that higher-level phenomena somehow would just go away when lower-level phenomena became better known.

Yet the siren song of reductionism will ever, it seems, be heard. Ultra-Darwinians are not molecular biologists. They were, for the most part, all trained in the population genetics tradition, and those molecular biologists who now claim that theirs is the key to the truth annoy ultra-Darwinians no end! But ultra-Darwinians have responded very clearly to the molecular revolution. Ours is the information age, and ultra-Darwinians quite naturally see evolution as a matter of the fate of genetic information. Indeed, we all see evolution in those terms.

The ultra-Darwinian response to the molecular revolution has been to insist that there is a titanic struggle going on—among genes, among organisms, perhaps among groups—to leave more copies of their genetic information to the next generation. I have argued that this represents an inappropriate reversal of the true vector of natural selection. Darwin's original description saw that economic success biases reproductive success, and that such an effect inevitably biases the transmission of heritable features from one generation to the next.

Ultra-Darwinians have flipped the story around. Competition for economic resources only goes on because organisms—or (fide Richard Dawkins) their genes—are locked in a combative, competitive struggle to pass along as many copies of themselves as possible to the next generation. I have suggested that this insistence is part of a seemingly endless progression within evolutionary biology to see itself as a true science. Competitive transmission of genetic information has an active air about it, and science is about the things that exist and what they do. Seeing evolution as the mere passive accumulation, the scorecard recording what worked better than what in the economic arena, seems tame by comparison. And it is evidently not to the tastes of the ultra-Darwinian painters of the evolutionary landscape.

This admittedly rather unflattering picture of ultra-Darwinians sees them as classic reductionists who have tried to inject a pattern of determinative action to a field that is, by its very nature, historical. Their actions have led them to a striking anomaly—one which, I believe, has yet to be fully realized and appreciated. Ultra-Darwinians, led by George Williams, have made the undeniable point that natural selection cannot have "eyes" for the future—meaning that selection can only work on a generation-by-generation basis, recording what worked relatively better among a collection of genetically varying individuals. Selection benefits only organisms, not species.

But what have they replaced it with? Another claim, ostensibly about actions of the present, but one that has direct, conspicuous, and probably equally false claims about the future. Organisms, it is said, are locked in a continual struggle to leave relatively more copies of their genes to the next generation. But how, by doing so, are organisms benefited? The postulate makes a bit more sense when we think, as Dawkins asked us to do with his *The Selfish Gene*, that the competition is really among variant forms of the genes themselves. But either way, it is a postulate about the future, not the present. Ultra-Darwinians, so it seems to me, have replaced one shopworn, incorrect adage about the supposed ability of evolution to govern the future with another one. I don't see any more sense in the new version than in the old.

In a book that has made no bones of its single-sided advocacy—one intended to offset similar productions from the other side of the High Table—I have had rather scant praise for the ultra-Darwinians. (I have, I hope, revealed my general respect for the practitioners even as I have chastised them for persistently misunderstanding us, and for promulgating their own faulty doctrines.) Time now to acknowledge what I take to be the main core of good in ultra-Darwinian works; for therein lies the sharpest, most succinct path to revealing our shared vision, as well as the distinctions that divide us naturalists from the ultra-Darwinian camp.

The ultra-Darwinian bailiwick—its base camp—is population genetics. Describing the dynamics of changes in gene frequency under all manner of conditions is the natural province of this branch of evolutionary biology, and progress continues to be made. A reductionist—an ultra-Darwinian—tends to claim that all we need to know about evolution resides here. A naturalist, convinced that other generally larger-scale levels (entities) exist and play important roles in evolution, nonetheless does not deny the existence, or indeed the importance, of population genetics-level processes. That is, after all, the very province of natural selection and genetic drift.

The two traditions of painting the evolutionary landscape are far from symmetrical. Ultra-Darwinians and their immediate predecessors see the existence of genes, chromosomes, organisms, and populations. Rarely do they take the existence of larger-scale entities—species and ecosystems—seriously, or view them as anything more than simple epiphenomena borne out of competition for reproductive success. Naturalists, in upholding the existence and significance of such larger-scale entities, do not reciprocate. To deny the existence or importance of populations would be idle, petulant stupidity.

The result is that we can, and indeed do, appreciate the positive results achieved by ultra-Darwinians when they are applying their concepts and analytic tools to the problems for which they are appropriate. When, for example, the issue is the structure, function, and evolution of a peculiar form of breeding system, their work is often elegant, and their presuppositions are ideally suited to the task at hand.

It is their penchant for extending a reproductive perspective progressively farther afield from population-level reproductive phenomena that gets them into trouble. Take species: variously regarded as of no special significance (Williams) or as outgrowths of reproductive or economic competition, ultra-Darwinians generally don't see species for what (it seems to me) they really are: simple outgrowths of sexual reproduction. Not competitive reproduction, but just reproduction. Species are reservoirs of genetic information. They do not themselves play economic roles in natural systems, nor do they (as entire entities) compete in any real sense with one another, economically or reproductively. They just exist, but act as reservoirs for the genetic information that everyone (seemingly) agrees is what evolution is all about. Ultra-Darwinians are even further off base when they try to depict the inner workings of multispecies (avatars) ecosystems as simple, direct manifestations of within-species competition for reproductive success.

It comes down to pictures. We naturalists, I am convinced, offer a less assumption ridden, theory laden description of the nature of all manner of biological systems. We absorb what the molecular geneticists have to say about the structure and inner workings of the genome itself. Likewise, as the early architects of the evolutionary synthesis taught us to, we recognize the nature and significance of population-level phenomena. But we also offer descriptions of larger-scale systems: species and higher taxa, avatars, and ecosystems of progressively larger scale. We see these entities as simple outcomes of the dual fact of organismic life: economics and reproduction.

It comes down to this: the competing allure of an essentially reductionist stance—with its charms of apparent simplicity and elegance—versus a partitioning of complexity into component systems—the naturalistic theoretical edifice that, while perhaps not as neat, seems to me a more accurate description of actual biological systems. Neither side of the High Table will end up convincing the other, nor does it seem likely that succeeding generations of High Table participants will come to any general agreement. But it is vital that the dialogue keep going. We are all trying to paint a better picture of Nature.

Bibliography

High Table discussion is carried on in many different venues and media, including scientific conferences, phone calls, faxes, letters, and even in genuine dinner table (and pub) conversations. But it is the published record that codifies those discussions. Perhaps more than any other scientific discipline, books continue to play a major role in the development of evolutionary theory. I list here the books referred to in the text, including all those that have had a critical impact especially in the years since 1959. I include as well all scientific articles mentioned in the text, plus a few others contributed by important discussants at the High Table who have not, at least as yet, written a book.

Darwin, C. 1859. *On the Origin of Species*. London: John Murray.
———. 1871. *The Descent of Man*. London: John Murray.
Dawkins, R. 1976. *The Selfish Gene*. New York: Oxford University Press.
———. 1982. *The Extended Phenotype: The Gene as the Unit of Selection*. San Francisco: W. H. Freeman and Co.
———. 1986. *The Blind Watchmaker*. New York: W. W. Norton.
Dobzhansky, T. 1937c; reprint 1982. *Genetics and the Origin of Species*. New York: Columbia University Press.
Eldredge, N. 1971. "The allopatric model and phylogeny in Paleozoic invertebrates." *Evolution* 25:156–167.

————. 1982. *The Monkey Business: A Scientist Looks at Creationism.* New York: Washington Square Press.

————. 1985a. *Unfinished Synthesis: Biological Hierarchies and Modern Evolutionary Thought.* New York: Oxford University Press.

————. 1985b. *Time Frames.* New York: Simon and Schuster.

————. 1986. "Information, economics and evolution." *Annual Reviews of Ecology and Systematics* 17:351–369.

————. 1989. *Macroevolutionary Dynamics: Species, Niches and Adaptive Peaks.* New York: McGraw-Hill.

Eldredge, N., and J. Cracraft. 1980. *Phylogenetic Patterns and the Evolutionary Process: Method and Theory in Comparative Biology.* New York: Columbia University Press.

Eldredge, N., and S. J. Gould. 1972. "Punctuated equilibria: an alternative to phyletic gradualism." In *Models in Paleobiology,* ed. T. J. M. Schopf, 82–115. San Francisco: Freeman, Cooper.

Eldredge, N., and M. Grene. 1992. *Interactions: The Biological Context of Social Systems.* New York: Columbia University Press.

Eldredge, N., and S. N. Salthe. 1984. "Hierarchy and evolution." *Oxford Surveys in Evolutionary Biology* 1:182–206.

Fisher, H. 1982. *The Sex Contract.* New York: W. Morrow.

Fisher, R. A. 1930; reprint 1958. *The Genetical Theory of Natural Selection.* New York: Dover.

Ghiselin, M. T. 1974. "A radical solution to the species problem." *Systematic Zoology* 23:536–544.

————. 1987. "Species concepts, individuality, and objectivity." *Biology and Philosophy* 2:127–143.

Gingerich, P. D. 1974. "Stratigraphic record of Early Eocene *Hyopsodus* and the geometry of mammalian phylogeny." *Nature* 248:107–109.

————. 1976. "Paleontology and phylogeny: patterns of evolution at the species level in Early Tertiary mammals." *American Journal of Science* 276:1–28.

Gleick, J. 1987. *CHAOS: Making a New Science.* New York: Viking Penquin.

Goldschmidt, R. 1940. *The Material Basis of Evolution.* New Haven, Conn.: Yale University Press.

Gould, S. J. 1977. *Ontogeny and Phylogeny.* Cambridge, Mass.: Harvard University Press.

————. 1980. "Is a new and general theory of evolution emerging?" *Paleobiology* 6:119–130.

————. 1982. "Darwinism and the expansion of evolutionary theory." *Science* 216:380–387.

————. *Wonderful Life: The Burgess Shale and the Nature of History.* New York: W. W. Norton.

Gould, S. J., and N. Eldredge. 1977. "Punctuated equilibria: the tempo and mode of evolution reconsidered." *Paleobiology* 3:115–151.

————. 1988a. "Punctuated equilibrium prevails." *Nature* 332:211–212.

————. 1988b. "Species selection: its range and power." *Nature* 334:19.

————. 1993. "Punctuated equilibrium comes of age." *Nature* 366:223–227.

Gould, S. J., and R. C. Lewontin. 1979. "The spandrels of San Marco and the Panglossian paradigm: a critique of the adaptationist programme." *Proceedings of the Royal Society, London* 205:581–598.

Gould, S. J., and E. S. Vrba. 1982. "Exaptation—a missing term in the science of form." *Paleobiology* 8:4–15.

Greenacre, M. J., and E. S. Vrba. 1984. "A correspondence analysis of biological census data." *Ecology* 65:984–997.

Grene, M. 1959. "Two evolutionary theories." *British Journal for the Philosophy of Science* 9:110–127, 185–193.

Hamilton, W. D. 1964a. "The genetical evolution of social behavior, I." *Journal of Theoretical Biology* 7:1–16.

————. 1964b. "The genetical evolution of social behavior, II." *Journal of Theoretical Biology* 7:17–52.

Hennig, W. 1966. *Phylogenetic Systematics*. Urbana: University of Illinois Press.

Hull, D. L. 1973. *Darwin and His Critics*. Cambridge, Mass.: Harvard University Press.

————. 1976. "Are species really individuals?" *Systematic Zoology* 25:174–191.

Kauffman, S. A. 1993. *The Origins of Order: Self-Organization and Selection in Evolution*. New York: Oxford University Press.

Kricher, John C. 1989. *A Neotropical Companion*. Princeton, N.J.: Princeton University Press.

Kuhn, T. S. 1962. *The Structure of Scientific Revolutions*. Chicago: University of Chicago Press.

Lande, R. 1976. "Natural selection and random genetic drift in phenotypic evolution." *Evolution* 30:314–334.

————. 1986. "The dynamics of peak shifts and the pattern of morphological evolution." *Paleobiology* 12:343–354.

Lauder, G. V. 1986. "Homology, analogy and the evolution of behavior." In *The Evolution of Behavior*, M. Nitecki and J. Kitchell, eds. pp. 9–40. Chicago: University of Chicago Press.

Lewontin, R. C. 1978. "Adaptation." *Scientific American* 239:212–230.

Maynard Smith, J. 1975. *The Theory of Evolution*, 3rd ed. Harmondsworth, England: Penguin Books.

————. 1978. *The Evolution of Sex*. Cambridge: Cambridge University Press.

————. 1984. "Palaeontology at the high table." *Nature* 309:401–402.

————. 1987. "Darwinism stays unpunctured." *Nature* 330:516.

————. 1988. "Punctuation in perspective." *Nature* 332:311–312.

Mayr, E. 1942; reprint 1982. *Systematics and the Origin of Species*. New York: Columbia University Press.

————. 1961. "Cause and effect in biology." *Science* 134:1501–1506.

————. 1963. *Animal Species and Evolution*. Cambridge, Mass.: Harvard University Press.

————. 1982. *The Growth of Biological Thought.* Cambridge, Mass.: Harvard University Press.

Michod, R. E., and B. R. Levin, eds. 1988. *The Evolution of Sex.* Sunderland, Mass.: Sinauer Associates.

Paterson, H. E. H. 1985. "The recognition concept of species." In ed. E. S. Vrba *Species and Speciation, Transvaal Museum Monographs* 4:21–29.

Schankler, D. M. 1981. "Local extinction and ecological re-entry of early Eocene mammals." *Nature* 293:135–138.

Sheldon, P. 1987. "Parallel gradualistic evolution of Ordovician trilobites." *Nature* 330:561–563.

Simpson, G. G. 1944. *Tempo and Mode in Evolution.* New York: Columbia University Press.

————. 1953. *The Major Features of Evolution.* New York: Columbia University Press.

————. 1961. *Principles of Animal Taxonomy.* New York: Columbia University Press.

Stanley, S. M. 1975. "A theory of evolution above the species level." *Proceedings of the National Academy of Sciences* 72:646–650.

————. 1979. *Macroevolution: Pattern and Process.* San Francisco: W. H. Freeman.

Stenseth, N. C., and J. Maynard Smith. 1984. "Coevolution in ecosystems: Red Queen evolution or stasis?" *Evolution* 38:870–880.

Van Valen, L. 1973. "A new evolutionary law." *Evolutionary Theory* 1:1–30.

Vrba, E. S. 1980. "Evolution, species and fossils: How does life evolve?" *South African Journal of Science* 76:61–84.

————. 1984a. "What is species selection?" *Systematic Zoology* 33:318–328.

————. 1984b. "Evolutionary pattern and process in the sister-group Alcelaphini-Aepycerotini (Mammalia:Bovidae)." In *Living Fossils,* eds. N. Eldredge and S. M. Stanley, 62–79. New York: Springer-Verlag.

————. 1985. "Environment and evolution: alternative causes of the temporal distribution of evolutionary events." *South African Journal of Science* 81:229–236.

Vrba, E. S., and N. Eldredge. 1984. "Individuals, hierarchies and processes: towards a more complete evolutionary theory." *Paleobiology* 10:146–171.

Vrba, E. S., and S. J. Gould. 1986. "The hierarchical expansion of sorting and selection: sorting and selection cannot be equated." *Paleobiology* 12:217–228.

Whewell, W. 1837. *History of the Inductive Sciences.* London: Parker.

Williams, G. C. 1966. *Adaptation and Natural Selection: A Critique of Some Current Evolutionary Thought.* Princeton, N.J.: Princeton University Press.

————. 1975. *Sex and Evolution.* Princeton, N.J.: Princeton University Press.

————. 1992. *Natural Selection. Domains, Levels, and Applications.* New York: Oxford University Press.

Wilson, E. O. 1975. *Sociobiology.* Cambridge,Mass.: Harvard University Press.

————. 1985. "The sociogenesis of insect colonies." *Science* 228:1489–1495.

Woolfenden, G. E., and J. W. Fitzpatrick. 1984. *The Florida Scrub Jay: Demography of a Cooperative-Breeding Bird.* Princeton, N.J.: Princeton University Press.

Wright, S. 1931. "Evolution in Mendelian populations." *Genetics* 16:97–159.

———. 1932. "The roles of mutation, inbreeding, crossbreeding, and selection in evolution." *Proceedings of the Sixth International Congress of Genetics* 1:356–366.

———. 1982. "Character change, speciation, and the higher taxa." *Evolution* 36:427–443.

Wynne-Edwards, V. C. 1962. *Animal Dispersion in Relation to Social Behavior.* New York: Hafner.

———. 1965. "Self-regulating systems in populations of animals." *Science* 147:1543–1548.

———. 1986. *Evolution Through Group Selection.* Boston: Blackwell Scientific Publications.

Index